Wh ... *the*
Greer ... Trucks

CW01650412

"...an extraorc ... *le to Cars and Trucks* will change ... ow have the ability to make a real difference in the health of our fragile planet."

—JACK GILLIS, Director of Public Affairs, Consumer Federation of America, and author of *The Car Book* and *The Truck, Van and 4x4 Book*

"What's novel about the *Guide* is that it doesn't rely on the traditional miles-per-gallon rating to determine what's green. Instead the criteria are: Fuel costs per year...Health costs per year...Tons of greenhouse gases emitted per year."

—MSNBC News

"The *Green Guide to Cars and Trucks* helps you avoid the gas guzzlers and choose a cleaner, more efficient car."

—CARL POPE, Executive Director, Sierra Club

"Use it when shopping for a new car!"

—JOAN CLAYBROOK, Executive Director, Public Citizen

"Don't go after your next vehicle without first consulting the *Green Guide to Cars and Trucks*!"

—*THE MIDWEST BOOK REVIEW*'s "Internet Bookwatch"

"If we're going to buy cars, we might as well buy the ones that are the most environmentally friendly. The *Green Guide to Cars and Trucks* is the best resource available for helping us do that."

—JOEL MAKOWER, *CO-OP AMERICA QUARTERLY*

"Unhealthy air, global warming, oil dependence—all of these problems are linked to our love affair with the automobile. ACEEE's *Green Guide to Cars and Trucks* gives consumers the information they need to make smart choices that do less harm to the environment."

—JOHN ADAMS, Executive Director, Natural Resources Defense Council

"...practical buying advice, and a good primer on the eco-impacts of vehicles on global warming and public health."

—*THE GREEN BUSINESS LETTER*

"If you care about the environment and are about to buy a new vehicle, this is the book for you."

—HOWARD RIS, Executive Director, Union of Concerned Scientists

"If you are about to purchase a...car or truck,...check out this book before you drive off the lot."

—South Coast Air Quality Management District's *ADVISOR*

"It will help you find a cleaner, greener vehicle for the driving you can't avoid."

—MICHAEL REPLOGLE, Federal Transportation Director, Environmental Defense Fund

"The *Green Guide to Cars and Trucks* will help people pick cars that are easy on the environment and on the pocketbook when folks go to the gas pump."

—HANK DITTMAR, Surface Transportation Policy Project

"This book is well done and is chock-full of meaningful, accessible information."

—Campaign on Auto Pollution's *GETTING THERE*

"If you insist on driving a car, check out the *Green Guide to Cars and Trucks*."

—Earth Island Institute's *JOURNAL*

Green Guide to Cars and Trucks

Green Guide to Cars and Trucks

Model Year 1999

by John DeCicco
and Martin Thomas

American Council for an Energy-Efficient Economy
Washington, D.C.
1999

Green Guide to Cars and Trucks: Model Year 1999

Copyright © 1999 by the American Council for an Energy-Efficient Economy

Reproduction in whole or in part is strictly prohibited without prior written approval of the American Council for an Energy-Efficient Economy (ACEEE), except that reasonable portions may be reproduced or quoted as part of a review or other story about this publication. All rights reserved.

Published by the American Council for an Energy-Efficient Economy, 1001 Connecticut Avenue, N.W., Suite 801, Washington, D.C. 20036

Printed in the United States of America

NOTICE

This publication was prepared by the American Council for an Energy-Efficient Economy (ACEEE). Neither ACEEE, the authors of this publication, nor any person acting on behalf of ACEEE: (a) makes any warranty, expressed or implied, with respect to the use of any information disclosed in this publication; (b) guarantees that such use does not infringe privately owned rights; (c) assumes any liabilities with respect to the use of, or for damages resulting from the use of, any information disclosed in this publication. ACEEE neither expresses nor implies any endorsement or recommendation, for or against, any of the products discussed in this publication. The statistics, ratings, and other data disclosed here are provided strictly for the purpose of providing information to the public and readers are left to make their own judgments regarding use of the information.

This publication was prepared on the basis of publicly available data and every attempt was made to insure that the data were current and accurate when the publication went to press. However, new or updated data and other information released by automakers, government agencies, and researchers may affect the information presented. ACEEE welcomes new information which might be reflected in future editions of this publication. New or updated information on vehicles covered here, or on vehicles not covered here for which information subsequently becomes available, will be posted on our website:

http://aceee.org/greenercars

Written comments on this publication are welcome; send your comments to:

attn: Automotive Editor
mail: American Council for an Energy-Efficient Economy
1001 Connecticut Avenue N.W., Suite 801
Washington, D.C. 20036
e-mail: ace3pubs@ix.netcom.com

Printed on recycled paper

Table of Contents

Acknowledgments

The authors would like to acknowledge the many individuals whose advice and suggestions helped us launch this publication last year and improve it this year. For technical comments and other assistance, we thank Jeff Alson, Jane Armstrong, Cheryl Bynum, Meredith Duran, Robert French, Tracie Jackson, Cindy Jacobs, Chris Lieske, Phil Lorang, Jim Markey, Terry Newell, Seung Park, Lois Platte, Lisa Snapp, and Arman Tanman of the U.S. Environmental Protection Agency (with special gratitude to Seung Park for his assistance in accessing the data used to prepare the guide) and Robert Cross and Paul Hughes of the California Air Resources Board. We are also grateful to Hiu Au, Rich Bechtold, Dan Becker, Mark Delucchi, Howard Geller, John German, Jack Gillis, Marc Goldsmith, Charles Griffith, Dan Gustafson, Roland Hwang, Greg Keoleian, Jason Mark, Marc Ross, Rick Tempchin, Michael Walsh, Michael Wang, and Tom Wenzel. We thank representatives of BMW, Chrysler, Daimler Benz, Ford, General Motors, Honda, Hyundai, Isuzu, Nissan, Saab, Subaru, Toyota, Volkswagen, and Volvo for providing data and comments.

At ACEEE, we thank Eric Stragar and Glee Murray for handling production and Renee Nida for proofreading. We are grateful to the U.S. Environmental Protection Agency's Office of Mobile Sources for support of research and methodology development as well as technical assistance. For supporting the publication of the *Green Guide to Cars and Trucks*, we express our gratitude to the Compton Foundation, the Changing Horizons Charitable Trust, The Energy Foundation, the Joyce Foundation, and Alida R. Messinger. Finally, since this edition builds on the original, we remain appreciative of all who were acknowledged last year as well.

Introduction

Many considerations go into buying a new car or light truck. You'll consider price, styling, comfort, performance, safety, reliability, and of course, how well the vehicle will serve your needs. The decision comes down to cost versus value: how much you are willing to pay for the features you want to get. But the costs of car use go beyond what's on the sticker and what you'll spend on fuel and repairs. There are hidden but very real environmental costs, due to adverse health impacts of air pollution, oil spills and fouling of water supplies, damage to habitats, and the growing risks of climate disruption. If you care about the future of our environment, then what you value goes beyond performance or styling and the options packages featured in the showroom.

The *Green Guide to Cars and Trucks* will help you choose a greener (cleaner and more fuel-efficient) vehicle, one that minimizes harm to the environment while meeting your transportation needs.

- For families with several vehicles, the pollution from their cars, vans, sport utilities, and pickup trucks is often greater than that from electricity and heating fuel use, waste disposal, and other household activities.
- Automobile pollution can be more dangerous than similar amounts of pollution from large sources such as power plants, since car and truck emissions are quite literally "in your face" where we live, work, shop, and play.
- Much more will have to be done to clean up cars and light trucks in order to avoid pollution alerts and bring air quality up to more healthful levels in many U.S. cities and regions.

Car and truck pollution not only harms our health in the present, but also contributes to global warming, bringing greater problems in years to come. A watchword of environmentalism is the concept of "sustainability." An action is sustainable if it serves our needs today without jeopardizing the ability of future generations (our children and grandchildren) to meet their needs. The large amounts of energy consumed and pollution produced by cars and trucks are among the reasons our transportation system is not sustainable.

Carbon dioxide (CO_2) is the most important greenhouse gas (GHG), referring to substances that trap heat in the earth's atmosphere and cause global warming. The largest portion of harmful GHG emissions is the CO_2 released from burning fossil fuels: oil, coal, and natural gas.

Carbon dioxide (CO_2) emissions from the consumption of gasoline, diesel, and other fossil fuels are the principal cause of global warming, which brings health and economic risks from climate change.

- The United States is the world's largest emitter of CO_2 and our emissions are still rising steadily. It will take decades before China and other rapidly growing economies reach our levels of CO_2 pollution per person.
- American cars and light trucks alone account for more fossil fuel CO_2 emissions than the total nationwide emissions of all but three other countries in the world.
- Although many measures will be needed both here and abroad, improving the fuel efficiency of cars and light trucks is one of the single biggest steps we can take to reduce global warming.

Making motor vehicles much cleaner and more efficient is an important step toward sustainable transportation. A large part of this task is up to automakers, but choosing a greener vehicle is a step you can take that will head us in the right direction.

It is clear that automobiles—this book's general term for cars, pickup trucks, station wagons, vans, sport utilities, and other personal motor vehicles—are among our largest causes of environmental damage. As shown in the adjoining illustration, environmental impacts start when automobiles are made, continue throughout their life on the road, and don't end even when the vehicle is scrapped, since waste disposal creates pollution, too.

Even the cleanest and most efficient vehicle on the market today still pollutes the air and otherwise damages the environment. A number of air pollutants are associated with automobiles:

- Fine airborne particulate matter (PM) causes lung trouble—shortness of breath, worsening of respiratory diseases and heart conditions, lung damage, and cancer.
- Nitrogen oxides (NO_x) aggravate respiratory problems, both directly and indirectly, by forming PM and smog; NO_x also causes acid rain and damages aquatic environments.

The Environmental Impacts of Automobiles

Ecological concerns start in the factories that manufacture a car and the parts and materials that go into it; workers and their communities are exposed to air and water pollution, toxins, and other hazards.

Gasoline and diesel fuel are poisonous to humans, plants and animals, and their fumes are toxic. Air and water pollution occur at oil wells, refineries, and filling stations and when petroleum products are transported by ship, pipeline, and truck. The more fuel we burn, the greater the fuel-supply impacts.

Much of the pollution from today's average automobile is in the exhaust. Even though government regulations prevent properly functioning new cars from polluting as much as they did in the past, personal vehicles still cause smog, toxics, and invisible but deadly fine particles that foul the air.

Whether in a crash or by old age, an automobile and its parts eventually get scrapped. Some components, including most of the steel, are recycled, but vehicle disposal still contaminates the air, water, and land.

Consumption of gasoline, diesel, and other petroleum products is the largest source of the greenhouse gas emissions that cause global warming. The risks from this growing form of global pollution include increased storms, droughts, and heat waves, spread of tropical diseases, rising sea levels,coastal flooding, and damage to agriculture, the economy, and ecosystems.

A gallon of gasoline weighs just over 6 pounds. When burned, the carbon in it combines with oxygen from the air to produce about 19 pounds of CO_2. But counting the energy that went into making and distributing the fuel, the total global warming impact equals 30 pounds of CO_2 emissions per gallon.

- Sulfur dioxide (SO_2) also irritates the lungs, and it contributes to forming PM as well as acid rain.
- Hydrocarbons (HC) are volatile organic compounds that cause smog and are toxic and carcinogenic.
- Carbon monoxide (CO) is a poisonous gas that impairs the flow of oxygen to the brain and other parts of the body.
- Carbon dioxide (CO_2) is not normally harmful, but as already noted, the huge amount of CO_2 released by burning gasoline and other fossil fuels is causing global warming.

Understanding the polluting effects of automobiles can help one appreciate the importance of considering a vehicle's greenness (or lack of greenness) when it comes time to purchase one. More information on why cars and trucks aren't ever truly "green" is given later in this book, under ***Automobiles and the Environment.***

Greener Transport Also Means Reducing Driving

A vehicle's greenness depends not only on its design, but also on how it is used. A car is greener when it's carrying two people rather than one and it's greener still with three. And it's greenest of all if left at home when there's a cleaner way to go: by foot or by bicycle, by bus or by train, and even by wire (as in telecommuting or videoconferencing). So, consider your opportunities to reduce car use when practical, by walking or biking for short trips, ridesharing, and combining several errands into one trip.

Our options for getting to work or school, shopping or recreation, conducting business, and visiting family or friends depend very much on where we live. Choosing where you live for its walkability and convenience to work, school, or transit stations—what planners call location efficiency—is a key opportunity to reduce your need for driving. In some areas, people find it surprisingly easy to do without a car at all.

Commuters may be eligible for benefits from their employers for transit or carpooling, or can receive cash by simply walking, biking, or telecommuting. See details on "Commuter Choice" programs on p. 90.

Buying Green

The *Green Guide to Cars and Trucks* will help you select a vehicle that does the least harm to the planet while meeting your transportation needs. Based on official emissions and fuel-economy test results and other specifications reported by auto manufacturers, we calculated a Green Score for each car, van, pickup, and sport utility on the market in 1999. The Green Score falls on a scale of zero to 100. A higher score implies a greener car, meaning a vehicle having lower environmental impacts.

Green Scores and Class Rankings

In our tables, vehicles are grouped together by class (that is, the type or body style, such as midsize car, minivan, standard pickup, and so on). To summarize our ratings and make it easy to find the top-rated vehicles in each class, we use five symbols based on a model's rank within its class. No vehicle gets a Superior (✔) rating if its Green Score is worse than the overall average of all vehicles offered this model year, even if it ranks among the best in its class. Look for models with a check mark (✔), or next best, an upward-pointing triangle (▲), to find those that have the greenest scores.

To make it easy to find the most environmentally friendly new cars and trucks available this year, the next section of the *Guide* features "The Best of '99" list. It provides abbreviated listings of the vehicles having the highest Green Scores in each class. The easiest way to buy green is to select a vehicle from the "Best of" listings.

Class Ranking Symbols

✔	Superior
▲	Above Average
○	Average
▽	Below Average
✖	Inferior

The symbols indicate a vehicle's environmental performance relative to others in its class.

Our master table provides detail on each vehicle's emissions standard, fuel economy, fuel costs, health effects, CO_2 emissions, and overall environmental impact, along with its Green Score and Class Ranking. If you have identified a set of

Different Versions of a Model Have Different Green Scores

Nearly all models come in different configurations, meaning different choices of engine, transmission, and other major options. A model may also be available in versions that meet different emissions standards or run on various alternative fuels. To match a model in the show room with a listing in the *Green Guide to Cars and Trucks*, first match its emissions standard. Then match its city and highway fuel economy ratings, which vary with engine size and transmission type.

models to consider, you can look them up in this table to comparison shop with the environment in mind.

For most consumers, the Green Score and Class Ranking provide a good indication of a car's environmental performance. We explain the factors behind the Green Score in the Appendix, "How the Ratings are Calculated." In general, a vehicle is greener if it is cleaner (having lower pollution) or more fuel efficient (resulting in lower energy consumption and CO_2 emissions).

All new vehicles for sale in the United States are certified to meet either Federal emissions standards, set by the U.S. Environmental Protection Agency (EPA), or California standards, set by the California Air Resources Board (CARB). These exhaust emissions standards limit the amounts of key pollutants coming from a vehicle's tailpipe and leaks in its fuel system. California has a set of standards that are progressively more stringent by pollutant. Except for a few makes and models, however, only Federally certified versions may be available in many states.

The principal tailpipe standards applicable to 1999 vehicles are:

Tier 1 — the prevailing Federal (EPA) standard.

Tier 1-D — the Federal diesel standard; permits higher NO_x emissions.

HDT — Heavy Duty Truck, the weakest standard for light trucks.

TLEV — Transitional Low Emission Vehicle, the weakest California standard.

LEV — Low Emission Vehicle, an intermediate California standard about twice as stringent as Tier 1.

ULEV — Ultra Low Emission Vehicle, a stronger California standard emphasizing very low HC emissions.

SULEV — Super Ultra Low Emission Vehicle, a California standard even tighter than ULEV and prohibiting emissions of fuel vapors.

ZEV Zero Emission Vehicle, a California standard prohibiting any tailpipe emissions.

The levels of pollution permitted under each standard depend on the vehicle type and many light trucks have weaker standards than cars. There are other standards not listed above, and new standards of increasing stringency are being developed by EPA and CARB. Each

How to Tell which Emission Standard a Vehicle Meets

In many states, most 1999 models meet only the Federal standards. California, Massachusetts, and New York are the main exceptions, where new vehicles must meet one of the California standards. California-certified automobiles will also be available in a number of other states. Look for vehicles bearing the label "Massachusetts Emissions Package" or "New York Emissions Package" elsewhere in the Northeast, for example. The labeling is better in California, where vehicles have an under-the-hood emissions label specifying which standard the vehicle meets, so just pop the hood and have a look. Some manufacturers are or will soon be offering California-certified vehicles nationwide, so ask your dealer if a less polluting model is available.

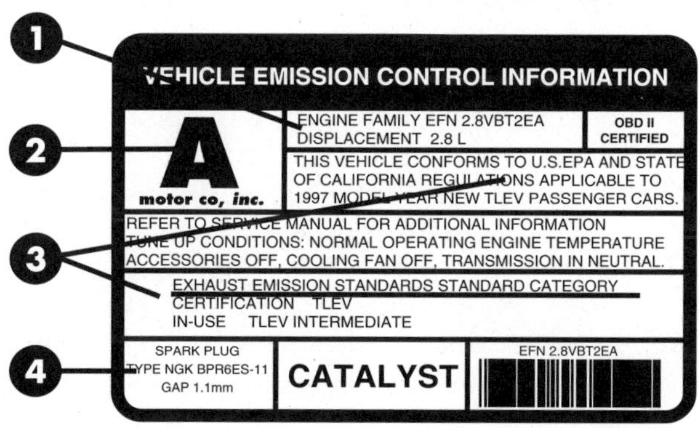

1 Engine family number
2 Manufacturer identification
3 Emissions standard identification
4 Tune-up information

Image credit: California Air Resources Board

Green Buyer Alert

The pollution coming from vehicles differs depending on the standards they meet, how well their emissions controls work in practice, how they are driven and maintained, and the quantity and quality of the fuel they burn. Vans, pickup trucks, and sport utilities are classified by the government as light trucks, and so have less stringent emissions and fuel economy standards than passenger cars. As a result, and as the ratings in this guide indicate, the average light truck pollutes more than the average car.

model's precise emissions standard is reflected in its Green Score and Class Ranking.

Automakers also rate their vehicles for fuel economy (miles per gallon—MPG) according to standard EPA tests of simulated city and highway driving. Unlike emissions standards, the fuel economy standards apply to manufacturers, rather than to individual vehicles. A manufacturer can sell models of varying fuel economy so long as their average fuel economy meets the standard. Light trucks (pickups, vans, and sport utilities) have a lower standard than passenger cars and station wagons. A new vehicle's sales sticker shows its city and highway MPG ratings; check these sticker MPG values to help you match a vehicle on the lot to a model listed in the *Green Guide to Cars and Trucks* tables.

Higher fuel economy means lower fuel consumption, savings on fuel costs, and reduced greenhouse gas emissions. Even in new vehicles, tailpipe standards don't fully control pollution in real-world driving conditions and don't capture the pollution effects related to the amount of fuel consumed. Therefore, among models meeting a given tailpipe emissions standard, higher fuel economy means lower total emissions of other pollutants as well.

Depending on where you live and how you use the vehicle, you may want to give greater or lesser importance to some factors over others. For this reason, our master table also provides details on a vehicle's health impacts, its global warming impacts, and its likely fuel costs. Buyers in urban areas, particularly those with serious air quality problems, should emphasize models that meet tighter tailpipe standards, such as the California standards noted above.

The Best of '99

The easiest way to choose a clean and efficient vehicle is to use this section of the *Guide*, which lists the greenest models in each vehicle class. If, for example, you are in the market for a minivan, look under Minivans to find the top-rated models. Also look to see if you can find a greener score among models in a similar class. For example, Minivan shoppers might want to check out Midsize Wagons.

We separately list the best Federally certified and California-certified vehicles so you can locate the models that will be available in your state or region. Some California-certified Low Emission Vehicles (LEVs), such as Honda Civics and Accords, are available nationwide.

If you are considering a model that is made in a California-certified version, ask your dealer for it even if you are not in California, Massachusetts, or New York. Some of these models might be available elsewhere.

Even if you can only obtain a Tier 1 vehicle this year, asking for cleaner cars provides feedback to the automakers, and cleaner versions might become more widely available in the future.

Our "Best of '99" tables list only the general vehicle make and model, not all of its configurations. See the master table starting on page 27 for complete information.

Even New Vehicles Aren't as Clean as You Think

Automakers often claim that their new cars are "96 percent cleaner" than those of decades past. This statement is false. Just because a new car or truck meets the new tailpipe standards doesn't mean that its pollution is adequately regulated. In fact, actual emissions over the life of today's automobiles average 2 to 4 times higher than the standard levels. The automakers' claim is based on laboratory tests that have a poor relation to typical driving. After all, what counts for our health is the air we breathe in real-world traffic conditions, not in a laboratory with carefully controlled conditions. Our Green Scores account for this discrepancy between lab tests and real-world driving.

The Top-Rated Models by Vehicle Class

	Specifications*	Emission Standard	Fuel Economy City	Fuel Economy Hwy	Green Score	Class Ranking
TWO SEATERS						
AUTOMATICS						
California Certified						
GM EV-1	Electric[1]	ZEV	3.3	4.0	**57**	✔
MAZDA MX-5 MIATA	1.8L 4, auto	TLEV	23	28	**25**	▲
SUBCOMPACT CARS						
AUTOMATICS						
California Certified						
HONDA CIVIC GX	1.6L 4, auto [CNG][2]	ULEV	28	34	**46**	✔
CHEVROLET METRO†	1.3L 4, auto	LEV	30	34	**33**	▲
SUZUKI SWIFT†	1.3L 4, auto	LEV	30	34	**33**	▲
MITSUBISHI MIRAGE	1.5L 4, auto	LEV	28	36	**32**	▲
HONDA CIVIC	1.6L 4, auto	LEV★	28	35	**31**	▲
Federally Certified						
HONDA CIVIC HX	1.6L 4, CVT	Tier 1	34	38	**31**	▲
CHEVROLET METRO†	1.3L 4, auto	Tier 1	30	34	**30**	▲
SUZUKI SWIFT†	1.3L 4, auto	Tier 1	30	34	**30**	▲
TOYOTA TERCEL	1.5L 4, auto	Tier 1	30	37	**30**	▲
MANUALS						
California Certified						
CHEVROLET METRO†	1.0L 3, manual	LEV	41	47	**38**	✔
SUZUKI SWIFT†	1.3L 4, manual	LEV	39	43	**37**	✔
MITSUBISHI MIRAGE	1.5L 4, manual	LEV	33	40	**34**	✔
HONDA CIVIC	1.6L 4, manual	LEV★	32	37	**33**	▲
Federally Certified						
CHEVROLET METRO†	1.0L 3, manual	Tier 1	41	47	**35**	✔
SUZUKI SWIFT†	1.3L 4, manual	Tier 1	39	43	**34**	✔
HONDA CIVIC HX	1.6L 4, manual	Tier 1	35	43	**32**	▲

* *The specifications include engine displacement in liters (L), number of cylinders, transmission type, and fuel type (e.g., P for premium); see the key on pages 32-33 for further details.*

[1] *Electric vehicle fuel economy is given in miles per kilowatt hour (mi/kWh).*

[2] *Compressed natural gas (CNG) fuel economy is given as gasoline-equivalent MPG.*

† *These vehicles are "twins"—the same base model carrying different names.*

†† *These vehicles are "triplets"—the same base model carrying different names.*

★*California certified vehicle available nationwide.*

Honda's technologically advanced engine and continuously variable automatic transmission give the Civic HX a best-in-class Green Score without sacrificing performance.

Photo: American Honda Motor Company

	Specifications*	Emission Standard	Fuel Economy City	Fuel Economy Hwy	Green Score	Class Ranking
COMPACT CARS						
AUTOMATICS						
California Certified						
HONDA EV PLUS	Electric[1]	ZEV	2.0	2.0	**42**	✔
SATURN SL	1.9L 4, auto	LEV	27	37	**31**	✔
MAZDA PROTEGE	1.6L 4, auto	ULEV	26	33	**31**	✔
Federally Certified						
TOYOTA COROLLA†	1.8L 4, auto	Tier 1	28	36	**29**	▲
CHEVROLET PRIZM†	1.8L 4, auto	Tier 1	28	36	**29**	▲
SATURN SL	1.9L 4, auto	Tier 1	27	37	**28**	▲
MANUALS						
California Certified						
SATURN SL	1.9L 4, manual	LEV	29	40	**33**	✔
MAZDA PROTEGE	1.6L 4, manual	ULEV	29	34	**32**	✔
DODGE NEON	2.0L 4, manual	LEV	28	39	**32**	✔
FORD ESCORT	2.0L 4, manual	LEV	28	37	**31**	▲

The Saturn SL's high fuel economy and LEV technology on California certified models make it a top choice among compact cars. *Photo: General Motors Corporation*

	Specifications*	Emission Standard	Fuel Economy City	Fuel Economy Hwy	Green Score	Class Ranking
COMPACT CARS (cont.)						
MANUALS (cont.)						
Federally Certified						
SATURN SL	1.9L 4, manual	Tier 1	29	40	**30**	▲
TOYOTA COROLLA†	1.8L 4, manual	Tier 1	31	38	**30**	▲
CHEVROLET PRIZM†	1.8L 4, manual	Tier 1	31	37	**30**	▲
DODGE NEON	2.0L 4, manual	Tier 1	28	39	**29**	▲
SMALL WAGONS						
AUTOMATICS						
California Certified						
SATURN SW	1.9L 4, auto	LEV	26	35	**30**	▲
FORD ESCORT WAGON†	2.0L 4, auto	LEV	25	34	**29**	▲
MERCURY TRACER WAGON†	2.0L 4, auto	LEV	25	34	**29**	▲
SUZUKI ESTEEM WAGON	1.6L 4, auto	TLEV	26	33	**27**	○

* *The specifications include engine displacement in liters (L), number of cylinders, transmission type, and fuel type (e.g., P for premium); see the key on pages 32-33 for further details.*

† *These vehicles are "twins"—the same base model carrying different names.*

	Specifications*	Emission Standard	Fuel Economy City	Fuel Economy Hwy	Green Score	Class Ranking
SMALL WAGONS (cont.)						
AUTOMATICS (cont.)						
Federally Certified						
SATURN SW	1.9L 4, auto	Tier 1	26	35	**27**	○
SUZUKI ESTEEM WAGON	1.6L 4, auto	Tier 1	26	33	**27**	○
FORD ESCORT WAGON†	2.0L 4, auto	Tier 1	25	34	**27**	○
MERCURY TRACER WAGON†	2.0L 4, auto	Tier 1	25	34	**27**	○
MANUALS						
California Certified						
SATURN SW	1.9L 4, manual	LEV	28	38	**32**	✔
FORD ESCORT WAGON†	2.0L 4, manual	LEV	28	37	**31**	✔
MERCURY TRACER WAGON†	2.0L 4, manual	LEV	28	37	**31**	✔
SUZUKI ESTEEM WAGON	1.6L 4, manual	TLEV	30	36	**29**	▲
Federally Certified						
SUZUKI ESTEEM WAGON	1.6L 4, manual	Tier 1	30	36	**29**	▲
SATURN SW	1.9L 4, manual	Tier 1	28	38	**29**	▲
FORD ESCORT WAGON†	2.0L 4, manual	Tier 1	28	37	**29**	○
MERCURY TRACER WAGON†	2.0L 4, manual	Tier 1	28	37	**29**	○

MIDSIZE CARS						
AUTOMATICS						
California Certified						
HONDA ACCORD	2.3L 4, auto	ULEV	23	30	**27**	▲
TOYOTA CAMRY	2.2L 4, auto	LEV	23	30	**26**	▲
CHEVROLET MALIBU†	2.4L 4, auto	LEV	22	30	**26**	▲
OLDSMOBILE CUTLASS†	2.4L 4, auto	LEV	22	30	**26**	▲
PLYMOUTH BREEZE	2.0L 4, auto	LEV	22	31	**26**	▲
MERCEDES E320	3.2L 6, auto [P]	LEV★	21	30	**24**	○
Federally Certified						
HONDA ACCORD	2.3L 4, auto	Tier 1	22	29	**24**	○
TOYOTA CAMRY	2.2L 4, auto	Tier 1	23	30	**24**	○
MAZDA 626	2.0L 4, auto	Tier 1	22	29	**24**	○
MANUALS						
California Certified						
DODGE STRATUS†	2.0L 4, manual	LEV	26	37	**30**	✔
PLYMOUTH BREEZE†	2.0L 4, manual	LEV	26	37	**30**	✔
HONDA ACCORD	2.3L 4, manual	LEV	25	31	**28**	✔

★*California certified vehicle available nationwide.*

With a 4-cylinder "VTEC" engine and sophisticated emissions control, this version of the Honda Accord is the first midsize car that meets ULEV standards when running on California's extra-clean reformulated gasoline. *Photo: General Motors Corporation*

	Specifications*	Emission Standard	Fuel Economy City	Hwy	Green Score	Class Ranking
MIDSIZE CARS (cont.)						
MANUALS (cont.)						
California Certified *(cont.)*						
MAZDA 626	2.0L 4, manual	TLEV	26	33	**27**	▲
MITSUBISHI GALANT	2.3L 4, manual	LEV	23	31	**27**	▲
Federally Certified						
DODGE STRATUS†	2.0L 4, manual	Tier 1	26	37	**27**	✔
PLYMOUTH BREEZE†	2.0L 4, manual	Tier 1	26	37	**27**	✔
MAZDA 626	2.0L 4, manual	Tier 1	26	33	**27**	▲
HONDA ACCORD	2.3L 4, manual	Tier 1	25	31	**26**	▲
MITSUBISHI GALANT	2.3L 4, manual	Tier 1	23	31	**25**	▲
MIDSIZE WAGONS						
AUTOMATICS						
California Certified						
NISSAN ALTRA	Electric[1]	ZEV	2.9	3.3	**52**	✔
SUBARU LEGACY WAGON	2.2L 4, auto	LEV	22	29	**26**	▲
MERCEDES E320 WAGON	3.2L6, auto [P]	LEV★	20	28	**23**	○

* *The specifications include engine displacement in liters (L), number of cylinders, transmission type, and fuel type (e.g., P for premium); see the key on pages 32-33 for further details.*

† *These vehicles are "twins"—the same base model carrying different names.*

[1] *Electric vehicle fuel economy is given in miles per kilowatt hour (mi/kWh).*

★*California certified vehicle available nationwide.*

Use of compressed natural gas enables the CNG version of the full-sized Ford Crown Victoria to meet ULEV standards, making it the top-rated large sedan in 1999. *Photo: Ford Motor Company*

	Specifications*	Emission Standard	Fuel Economy City	Fuel Economy Hwy	Green Score	Class Ranking
MIDSIZE WAGONS (cont.)						
AUTOMATICS (cont.)						
Federally Certified						
SUBARU LEGACY WAGON	2.2L 4, auto	Tier 1	22	29	**23**	○
VOLKSWAGEN PASSAT WAGON	1.8L 4, auto [P]	Tier 1	21	31	**23**	○
MANUALS						
California Certified						
SUBARU LEGACY WAGON	2.2L 4, manual	LEV	22	29	**26**	✔
VOLKSWAGEN PASSAT WAGON	1.8L 4, manual [P]	TLEV	23	32	**25**	▲
Federally Certified						
VOLKSWAGEN PASSAT WAGON	1.8L 4, manual [P]	Tier 1	23	32	**24**	▲

	Specifications*	Emission Standard	Fuel Economy City	Fuel Economy Hwy	Green Score	Class Ranking
LARGE CARS						
AUTOMATICS						
California Certified						
FORD CROWN VICTORIA	4.6L 8, auto [CNG][2]	ULEV	14	20	**29**	✔
TOYOTA AVALON	3.0L 6, auto	LEV	21	29	**25**	✔
CHRYSLER CONCORDE†	2.7L 6, auto	TLEV	21	30	**23**	▲
DODGE INTREPID†	2.7L 6, auto	TLEV	21	30	**23**	▲

[2] *Compressed natural gas (CNG) fuel economy is given as gasoline-equivalent MPG.*

The front wheel drive, 2.4L 4-cylinder engine versions of the Dodge Caravan, and its Plymouth Voyager twin, are the greenest gasoline-powered minivans. If it has California LEV certification, a 3.0L 6-cylinder model scores nearly as well. *Photo: DaimlerChrysler Corporation*

	Specifications*	Emission Standard	Fuel Economy City	Fuel Economy Hwy	Green Score	Class Ranking
LARGE CARS (cont.)						
AUTOMATICS (cont.)						
Federally Certified						
CHRYSLER CONCORDE†	2.7L 6, auto	Tier 1	21	30	**23**	▲
DODGE INTREPID†	2.7L 6, auto	Tier 1	21	30	**23**	▲
TOYOTA AVALON	3.0L 6, auto	Tier 1	21	29	**23**	▲
BUICK LESABRE	3.8L 6, auto	Tier 1	19	30	**22**	▲
MINIVANS						
AUTOMATICS						
California Certified						
CHRYSLER EPIC	Electric[1]	ZEV	1.5	1.4	**30**	✔
DODGE CARAVAN†	2.4L 4, auto	LEV	20	26	**20**	✔
PLYMOUTH VOYAGER†	2.4L 4, auto	LEV	20	26	**20**	✔
Federally Certified						
DODGE CARAVAN†	2.4L 4, auto	Tier 1	20	26	**18**	○
PLYMOUTH VOYAGER†	2.4L 4, auto	Tier 1	20	26	**18**	○

* *The specifications include engine displacement in liters (L), number of cylinders, transmission type, and fuel type (e.g., P for premium); see the key on pages 32-33 for further details.*

[1] *Electric vehicle fuel economy is given in miles per kilowatt hour (mi/kWh).*

† *These vehicles are "twins"—the same base model carrying different names.*

High fuel economy helps 2.2L 4-cylinder, 2-wheel drive versions of the Chevrolet S-10 and its twins the GMC Sonoma and Isuzu Hombre score well among compact pickups. LEV versions and an electric S-10 are also available in some areas. *Photo: General Motors Corporation*

	Specifications*	Emission Standard	Fuel Economy City	Fuel Economy Hwy	Green Score	Class Ranking
LARGE VANS						
AUTOMATICS						
California Certified						
DODGE B2500 VAN/WAGON	5.2L 8, auto [CNG][2]	SULEV	14	19	**26**	✔
FORD E250 ECONOLINE	5.4L 8, auto [CNG][2]	SULEV	12	15	**20**	✔
DODGE B1500 VAN	3.9L 6, auto	LEV	15	17	**14**	▲
Federally Certified						
VOLKSWAGEN EUROVAN	2.8L 6, auto	Tier 1	15	20	**13**	▲
CHEVROLET G1500/G2500†	4.3L 6, auto	Tier 1	15	19	**13**	○
GMC G1500/G2500 SAVANA†	4.3L 6, auto	Tier 1	15	19	**13**	○
COMPACT PICKUPS						
AUTOMATICS						
California Certified						
CHEVROLET S10 PICKUP	Electric[1]	ZEV	2.2	2.4	**46**	✔
FORD RANGER PICKUP	Electric[1]	ZEV	2.6	2.3	**42**	✔
FORD RANGER PICKUP†	2.5L 4, auto	LEV	20	25	**23**	▲
MAZDA B2500†	2.5L 4, auto	LEV	20	25	**23**	▲
CHEVROLET S10 PICKUP ††	2.2L 4, auto	LEV	19	26	**23**	▲

[2] *Compressed natural gas (CNG) fuel economy is given as gasoline-equivalent MPG.*

†† *These vehicles are "triplets"—the same base model carrying different names.*

	Specifications*	Emission Standard	Fuel Economy City	Fuel Economy Hwy	Green Score	Class Ranking
COMPACT PICKUPS (cont.)						
AUTOMATICS (cont.)						
California Certified *(cont.)*						
GMC SONOMA††	2.2L 4, auto	LEV	19	26	**23**	▲
ISUZU HOMBRE PICKUP††	2.2L 4, auto	LEV	19	26	**23**	▲
Federally Certified						
FORD RANGER PICKUP†	2.5L 4, auto	Tier 1	20	25	**21**	○
MAZDA B2500†	2.5L 4, auto	Tier 1	20	25	**21**	○
CHEVROLET S10 PICKUP††	2.2L 4, auto	Tier 1	19	26	**21**	○
GMC SONOMA††	2.2L 4, auto	Tier 1	19	26	**21**	○
ISUZU HOMBRE PICKUP††	2.2L 4, auto	Tier 1	19	26	**21**	○
MANUALS						
California Certified						
CHEVROLET S10 PICKUP††	2.2L 4, manual	LEV	23	29	**26**	▲
GMC SONOMA††	2.2L 4, manual	LEV	23	29	**26**	▲
ISUZU HOMBRE PICKUP††	2.2L 4, manual	LEV	23	29	**26**	▲
NISSAN FRONTIER TRUCK	2.4L 4, manual	LEV	22	26	**25**	▲
Federally Certified						
CHEVROLET S10 PICKUP††	2.2L 4, manual	Tier 1	23	29	**23**	▲
GMC SONOMA††	2.2L 4, manual	Tier 1	23	29	**23**	▲
ISUZU HOMBRE PICKUP††	2.2L 4, manual	Tier 1	23	29	**23**	▲
NISSAN FRONTIER TRUCK	2.4L 4, manual	Tier 1	22	26	**23**	▲
TOYOTA TACOMA	2.4L 4, manual	Tier 1	22	27	**23**	▲
STANDARD PICKUPS						
AUTOMATICS						
California Certified						
FORD F250 PICKUP	5.4L 8, auto [CNG]²	SULEV	12	15	**22**	✔
FORD F250 PICKUP	5.4L 8, auto [CNG]²	ULEV	12	15	**20**	✔
CHEVROLET C1500 SILVERADO†	4.3L 6, auto	LEV	16	20	**16**	▲
GMC C1500 SIERRA†	4.3L 6, auto	LEV	16	20	**16**	▲
Federally Certified						
MAZDA B4000	4.0L 6, auto	Tier 1	16	22	**18**	▲
CHEVROLET C1500 SILVERADO†	4.3L 6, auto	Tier 1	16	20	**15**	▲
GMC C1500 SIERRA†	4.3L 6, auto	Tier 1	16	20	**15**	▲

** The specifications include engine displacement in liters (L), number of cylinders, transmission type, and fuel type (e.g., P for premium); see the key on pages 32-33 for further details.*

² Compressed natural gas (CNG) fuel economy is given as gasoline-equivalent MPG.

† These vehicles are "twins"—the same base model carrying different names.

†† These vehicles are "triplets"—the same base model carrying different names.

Both the nationally available 2-wheel drive gasoline and electric versions of Toyota's Rav 4 top the list of greenest small SUVs. *Photo: Toyota Motor Sales, U.S.A. and Wieck Photo DataBase, Inc.*

	Specifications*	Emission Standard	Fuel Economy City	Fuel Economy Hwy	Green Score	Class Ranking
STANDARD PICKUPS (cont.)						
MANUALS						
California Certified						
GMC C1500 SIERRA†	4.3L 6, manual	LEV	17	23	**17**	▲
CHEVROLET C1500 SILVERADO†	4.3L 6, manual	LEV	17	23	**17**	▲
Federally Certified						
MAZDA B4000	4.0L 6, manual	Tier 1	17	22	**19**	▲
GMC C1500 SIERRA†	4.3L 6, manual	Tier 1	17	23	**16**	▲
CHEVROLET C1500 SILVERADO†	4.3L 6, manual	Tier 1	17	23	**16**	▲
SMALL UTILITIES						
AUTOMATICS						
California Certified						
TOYOTA RAV4	Electric[1]	ZEV	3.2	2.6	**50**	✔
SUBARU FORESTER	2.5L 4, auto	TLEV	21	26	**22**	○
Federally Certified						
TOYOTA RAV4	2.0L 4, auto	Tier 1	24	29	**25**	▲
SUZUKI VITARA 2-DOOR	1.6L 4, auto	Tier 1	25	27	**25**	▲

[1] *Electric vehicle fuel economy is given in miles per kilowatt hour (mi/kWh).*

	Specifications*	Emission Standard	Fuel Economy City	Fuel Economy Hwy	Green Score	Class Ranking
SMALL UTILITIES (cont.)						
MANUALS						
California Certified						
ISUZU AMIGO	2.2L 4, manual	LEV	21	24	**23**	○
SUBARU FORESTER	2.5L 4, manual	TLEV	21	27	**23**	○
Federally Certified						
CHEVROLET TRACKER CONVERTIBLE	1.6L 4, manual	Tier 1	25	28	**25**	▲
SUZUKI VITARA 2-DOOR	1.6L 4, manual	Tier 1	25	28	**25**	▲
MEDIUM UTILITIES						
AUTOMATICS						
California Certified						
JEEP CHEROKEE	2.5L 4, auto	TLEV	18	22	**19**	▲
MITSUBISHI MONTERO	3.0L 6, auto	LEV	18	21	**18**	▲
Federally Certified						
JEEP CHEROKEE	2.5L 4, auto	Tier 1	18	22	**19**	▲
TOYOTA 4RUNNER	2.7L 4, auto	Tier 1	20	24	**17**	○
MANUALS						
California Certified						
MITSUBISHI MONTERO	2.3L 4, manual	LEV	22	24	**24**	✔
ISUZU RODEO	2.2L 4, manual	LEV	21	24	**23**	✔
Federally Certified						
MITSUBISHI MONTERO/NATIVA	2.3L 4, manual	Tier 1	22	24	**22**	✔
JEEP CHEROKEE	2.5L 4, manual	Tier 1	21	25	**22**	✔
LARGE UTILITIES						
AUTOMATICS						
California Certified						
CHEVROLET C1500 TAHOE†	5.7L 8, auto	LEV	14	18	**13**	▲
GMC C1500 YUKON†	5.7L 8, auto	LEV	14	18	**13**	▲
FORD EXPEDITION	4.6L 8, auto	LEV★	13	18	**13**	▲
Federally Certified						
CHEVROLET C1500 TAHOE†	5.7L 8, auto	Tier 1	14	18	**12**	▲
GMC C1500 YUKON†	5.7L 8, auto	Tier 1	14	18	**12**	▲

* *The specifications include engine displacement in liters (L), number of cylinders, transmission type, and fuel type (e.g., P for premium); see the key on pages 32-33 for further details.*

† *These vehicles are "twins"—the same base model carrying different names.*

★*California certified vehicle available nationwide.*

Highlights of the Model Year

Model year 1999 brings both good news and bad news for the environment. A number of automakers, notably Ford and Honda, are increasing their nationwide sales of Low Emissions Vehicles (LEVs). This will speed progress in cutting urban smog. On the other hand, last year saw record sales of light trucks, led by the still raging popularity of large sport utility vehicles. Automakers continue to expand their offerings of these gas guzzlers, with upscale versions leading the charge. You can now buy a huge, leather-lined "utility" truck with any one of many luxury marquees. This is not good news for the environment, and it is especially bad news for global warming.

Electric vehicles remain the top-scoring vehicles on the market. Pushed by California's Zero Emissions Vehicle regulations, nearly all of the major automakers have electric vehicles for sale. EVs remain the cutting edge of vehicle technology. But the cost and range limitations of batteries continue to keep EVs from being a practical choice for most buyers. Natural gas powered vehicles also score very well. While not having quite the range and cost limitations of EVs, the need for access to natural gas fueling stations means that CNG cars and trucks are usually a better choice for fleet buyers than for the average consumer.

This year, the biggest step forward in car and light truck pollution reduction comes from Ford Motor Company. It has certified its Windstar minivans and all of its sport utility vehicles—the Ford Explorer and Mercury Mountaineer, the Ford Expedition and Lincoln Navigator—as Low Emission Vehicles throughout the country, not just where required in California and parts of the Northeast. Honda took the lead on nationwide availability of last year, and continues to make its LEV-certified Civics and Accords widely available as well. The greater availability of LEVs represents real progress in cutting smog-forming pollutants. However, the lack of fuel economy improvement, particularly in minivans and sport utilities, makes for ongoing trouble with global warming pollution.

Behind the scenes, moreover, a quiet revolution has been taking place in the quality of emissions control for conventional cars and trucks. In researching the *Green Guide to Cars and Trucks*, we examine the most recently available data on real-world emissions control performance. These data show that, on average, manufacturers are doing a better job of designing emissions controls that keep pollution low under a variety of in-use conditions. The California Air Resources Board (CARB) and U.S. Environmental Protection Agency (EPA) check up on how well auto pollution controls are working in practice. Such oversight pushes automakers to fix flaws in their emissions control systems. New cars and trucks still aren't as clean as some automakers advertise. But progress is being made. Increased scrutiny by EPA and CARB, as well as tighter standards across the board, such as those proposed this year by CARB, can cut even more pollution from ordinary vehicles as well as save consumers money through avoided repairs and more cost-effective emissions controls.

Model year 1999 again brings stagnation, even backsliding, in terms of fuel economy. Most of the auto industry's manufacturing and marketing skills are not being applied to help the environment. Instead, the trend continues toward larger, more luxurious, more powerful (and often more dangerous) vehicles in the light truck classes. While a few hundred electric vehicles might be sold, half the market—approaching eight million vehicles—are various versions of pickups, vans, and sport utilities. Largely displacing sales of more efficient cars, sport ute sales have more than quadrupled over the past decade. That's why average new vehicle fuel economy is now over 1 MPG lower than it was ten years ago.

Although SUVs and other light trucks are clearly popular, we shouldn't forget that popularity is influenced by advertising and sales strategies. Automaker's strategies are in turn influenced by government policies, and this situation is quite relevant to the light truck market. Vehicles classified as light trucks are held to lower fuel economy standards. They are exempt from the gas guzzler tax that applies to passenger cars, even though they are used mainly for passenger transportation. Moveover, many light trucks enjoy a tariff protection. The combination of these federal policies presents a powerful financial incentive for automakers to increase their light truck sales, and that is undoubtedly part of the reason for the sales strategies that have helped push these vehicles' popularity to the high levels we've seen in recent years.

But the market, of course, is a give and take between consumers and manufacturers. As a consumer who cares about the environment, your purchase decisions can influence the products that automakers offer. Buying green can help expand the market for cleaner and more efficient automobiles. Remember also that the average car or light truck is likely to be operational for a dozen or more years. Even if you don't keep the vehicle you buy in 1999 for more than a few years,

PRACTICAL PICKS— A SELECTION OF GASOLINE VEHICLES THAT SCORE WELL

Make and Model	Specifications	Emission Standard	Fuel Economy City	Fuel Economy Hwy	Green Score
HONDA CIVIC	1.6L 4, auto	LEV★	28	35	**31**
CIVIC HX	1.6L 4, CVT	Tier 1	34	38	**31**
MITSUBISHI MIRAGE	1.5L 4, auto	Tier 1–LEV	28	36	**29–32**
MAZDA PROTEGE	1.6L 4, auto	Tier 1–ULEV	26	33	**27–31**
CHEVROLET PRIZM † TOYOTA COROLLA †	1.8L 4, auto	Tier 1–LEV	28	36	**29–30**
SATURN SW WAGON	1.9L 4, auto	Tier 1–LEV	26	35	**27–30**
HONDA ACCORD	2.3L 4, auto	Tier 1–LEV★, ULEV	23	30	**24–27**
SUBARU LEGACY WAGON	2.2L 4, auto	Tier 1– LEV	22	29	**23–26**
DODGE CARAVAN †	2.4L 4, auto 2wd	Tier 1–LEV	20	26	**18–20**
PLYMOUTH VOYAGER †	3.0L 6, auto 2wd	Tier 1–LEV	19	26	**18–19**
FORD RANGER PICKUP † MAZDA B2500 †	2.5L 4, auto 2wd	Tier 1–LEV	20	25	**21–23**
CHEVROLET S10 PICKUP † GMC SONOMA † ISUZU HOMBRE PICKUP †	2.2L 4, manual 2wd 4.3L 6, manual 2wd	Tier 1–LEV Tier 1–LEV	23 17	29 23	**21–23** **19–21**
MITSUBISHI MONTERO	2.3L 4, manual 2wd 3.0L 6, auto 2wd	Tier 1–LEV LEV	22 18	24 21	**22–24** **18**
JEEP CHEROKEE	2.5L 4, auto 2wd	Tier 1–TLEV	18	22	**19**

† *These vehicles are "twins"—the same base model carrying different names.*
★ *California certified vehicle available nationwide.*

your choice will affect the options available for used car buyers. Don't put an extra gas guzzler on the road for years to come!

Suppose you want to buy green, but need more than a small car and are not ready for an electric vehicle. Our "Best of '99" table shows that there are greener choices in all segments of the market. We highlight some of those choices here in what we call "Practical Picks." This table is a selection of gasoline vehicles that score well this year, excluding the smallest small cars and small sport utilities. The car picks include only automatics, even though many manual transmission versions have higher fuel economy. The excluded models that score even better might be good choices for some buyers. The point of Practical Picks is that there are ways to buy green even in the mainstream market.

THE 12 GREENEST VEHICLES IN 1999

Make and Model	Specifications	Emission Standard	Fuel Economy City	Fuel Economy Hwy	Green Score	Class Ranking
GM EV-1	Electric	ZEV	3.3	4.0	**57**	✔
NISSAN ALTRA	Electric[1]	ZEV	2.9	3.3	**52**	✔
TOYOTA RAV4	Electric[1]	ZEV	3.2	2.6	**50**	✔
HONDA CIVIC GX	1.6L 4, auto [CNG][2]	ULEV	28	34	**46**	✔
CHEVROLET S-10	Electric[1]	ZEV	2.2	2.4	**46**	✔
HONDA EV PLUS	Electric[1]	ZEV	2.0	2.0	**42**	✔
FORD RANGER	Electric[1]	ZEV	2.6	2.3	**42**	✔
CHEVROLET METRO	1.0L 3, manual	LEV	41	47	**38**	✔
SUZUKI SWIFT	1.3L 4, manual	LEV	39	43	**37**	✔
MITSUBISHI MIRAGE	1.5L 4, manual	LEV	33	40	**34**	✔
HONDA CIVIC	1.6L 4, manual	LEV★	32	37	**33**	▲
SATURN SC/SL	1.9L 4, manual	LEV	29	40	**33**	▲

★*Honda Civic LEVs are available nationwide.*

[1] *Electric vehicle fuel economy is given in miles per kilowatt hour (mi/kWh).*

[2] *CNG fuel economy is given as gasoline-equivalent MPG.*

Why Do Electric Vehicles Score So Well?

The greenest vehicle of '99 is the General Motors EV-1, and electric vehicles top the list in every class in which they are offered. For many people, this is no surprise. But others raise the question, "Even though EVs have zero tailpipe emissions, they use electricity from power plants, so don't they just produce pollution elsewhere?" Yes, they do, but there are two main reasons why electric vehicles are less harmful than gasoline or diesel vehicles.

One is that a given amount of pollution from a power plant is generally much less damaging to health than a similar amount from a tailpipe. Pollution from vehicle tailpipes, emitted at ground level and directly in city streets, exposes more people to higher concentrations of pollutants than do power plants, which are generally remote from populated areas and have stacks that allow the pollution to disperse before it reaches ground level.

The other reason is that real-world emissions from gasoline vehicles are much higher than the levels set by the tailpipe standards. This problem occurs for a variety of reasons, as pointed out earlier under "Even New Vehicles Aren't as Clean as You Think" on page 9. Battery electric vehicles avoid the emissions control malfunction and degradation problems that plague gasoline vehicles; by comparison,

A superbly designed advanced body structure and an energy-efficient electric drivetrain give GM's EV-1 excellent performance along with the highest Green Score for model year 1999.

Photo: General Motors Corporation

emissions from power plants are more reliably regulated within their permitted levels.

For any type of vehicle, greenhouse gas emissions depend largely on its fuel efficiency; CO_2 and other greenhouse gas emissions are equally damaging no matter where they are emitted. Electric vehicles can be made with very efficient drivetrains—the EV-1 is a good example of efficient overall design, so its GHG emissions are also exceptionally low.

Today's electric vehicles depend solely on batteries, which bring higher costs and restricted driving ranges compared to gasoline vehicles. Thus, EVs are not yet practical for everyone. Advanced electric vehicle technologies such as hybrid powertrains or fuel cells hold great promise for the future, and even if they have an on-board fuel, their emissions can be made quite low. Toyota has been marketing an ultra-clean hybrid vehicle in Japan, and other new developments may soon follow.

The next few years could be an exciting time for greener automotive technologies, so stay tuned. Watch our website (http://aceee.org/greenercars), where we'll post Green Scores for new products that come on the market throughout the year.

Why Do Some Light Trucks Score So Poorly?

The "12 Worst Vehicles for the Environment" list shows vehicles with Green Scores below 10 in one or more version. The list is

THE 12 WORST VEHICLES FOR THE ENVIRONMENT IN 1999

Make and Model (listed alphabetically)	Specifications	Emission Standard	Fuel Economy City	Fuel Economy Hwy	Green Score	Class Ranking
CADILLAC ESCALADE	5.7L 8, auto 4wd	Tier 1-LEV	12	16	**8–10**	▽
CHEVY SILVERADO / GMC SIERRA C2500, K1500, K2500 Pickups	6.0L 8, 2 and 4wd models; 5.7L 8, 4wd models.	HDT-LEV	12–14	16–18	**8–12**	✖
CHEVY TAHOE / GMC YUKON, K1500	5.7L 8, auto 4wd;	LEV-Tier 1	12	16	**8–10**	▽
K1500 Diesel	6.5L 8, auto 4wd [D]	Tier 1-D	15	19	**8**	✖
CHEVROLET / GMC SUBURBAN	5.7L 8, auto 2 and 4wd	HDT-Tier1	14	18	**9–12**	▽
DODGE B1500 / B2500 VAN/WAGON	5.9L 8, auto 2wd	HDT-LEV	12	17	**8–12**	✖
DODGE RAM 1500 / 2500 PICKUP	5.9L 8, auto 4wd	HDT-LEV	12	13	**7–9**	✖
FERRARI 550 MARANE	5.5L 12, manual [P] 2wd	Tier 1	9	14	**9**	✖
FORD E150/E250 CLUB WAGON/ECONOLINE	5.4L 8, auto 2wd	HDT-Tier 1	12	17	**8**	✖
FORD EXPEDITION / LINCOLN NAVIGATOR	5.4L 8, auto 4wd	LEV	12	16	**9–10**	✖
FORD F250 PICKUP	5.4L 8, auto 4wd	HDT-Tier 1	12–13	15–17	**9–12**	✖
ROLLS ROYCE SILVER SPUR / BENTLEY BROOKLANDS	6.8L 8, auto 2wd	Tier 1	10	15	**9**	✖
TOYOTA LAND CRUISER / LEXUS LX 470	4.7L 8, auto 4wd	Tier 1	13	16	**9**	▽

[D] denotes diesel fuel; [P] denotes premium gasoline.

dominated by large sport utilities and pickups with 8-cylinder engines and often with 4-wheel drive. The two exceptions are both exotic sports/luxury cars. If large light trucks were used mainly for true heavy-duty hauling, fewer would be sold and their pollution would be less of a problem. But in recent years, automakers have been marketing these vehicles for passenger use, pushing their sales into the millions and creating one of the most rapidly growing sources of pollution and greenhouse gas emissions in the country.

Since our Green Scores reflect both tailpipe emissions and fuel economy, the high fuel consumption alone would suffice to push down the environmental ratings of many light trucks. Moreover, most such vehicles are also subject to more lenient tailpipe emissions standards, so the result is two big strikes against environmental performance. Even some trucks that meet LEV standards are dirtier than the worst new cars. The largest SUVs are so massive that they are classified as heavy trucks and are exempt from fuel economy laws and strict tailpipe pollution standards. The pickups and sport utilities topping our "12 Worst" list rely on

Ford equipped all of its SUVs and minivans with LEV emissions controls for 1999, significantly reducing their smog-forming pollution nationwide. The 2-wheel drive, 4.6L engine Ford Expeditions make our Best of '99 list of one of the greener ("least bad") large SUVs. But with a larger engine and 4-wheel drive, the Expedition still joins similar large SUVs on our list of Worst Vehicles for the Environment. *Photo: Ford Motor Company*

engine blocks and chassis designs that are largely decades old, with relatively minor updating from a technological point of view (even though the styling may be new). These vehicles provide their capacity through brute force. Similar capacity could be more cleanly and efficiently provided through smart design, but automakers aren't offering truck designs on a par with the best car technology. These light trucks are also quite pricey—practical working trucks need not be so loaded with luxuries and other amenities. Inefficient technology, coupled with the marketing of power-performance and off-road features that are rarely used makes these vehicles a poor value from an economic as well as an environmental point of view.

The chart on p. 28 compares a range of family-oriented vehicles on the market this year. A large "sport ute" pollutes more—and costs more—than a minivan, and it pollutes much, much more than a station wagon. Similar comparisons can be made among the models offered by other manufacturers. Check the *Green Guide to Cars and Trucks* tables to find models within any size class that avoid overpowered engines and unnecessary 4-wheel drive. To "buy green," think

EMISSIONS COMPARISON OF FAMILY-ORIENTED VEHICLES

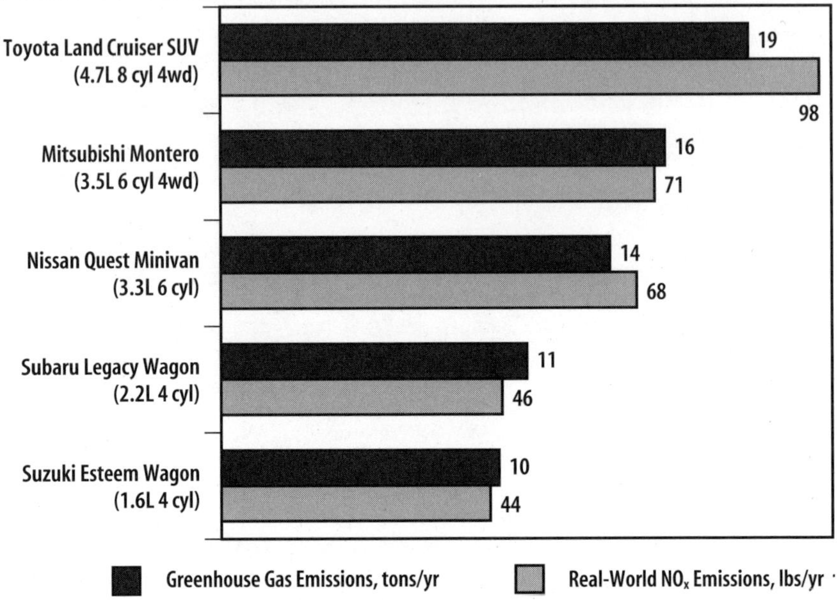

All models are Federally certified versions with automatic transmissions; pollution values are full-fuel-cycle estimates for 15,000 miles of annual driving.

carefully about the features that you really need in a vehicle, remembering that the best choice will be the greenest vehicle that meets your needs and fits your budget.

Bi- and Flex-Fuel Vehicles

A number of automakers are now offering vehicles with bi-fuel capability, meaning the ability to run on either an alternative fuel or gasoline. For example, some pickup trucks can be set to run on compressed natural gas (CNG) or gasoline, depending on what fuel is available to the driver. Flexible-fuel vehicles ("FFVs") can automatically detect the fuel put into the tank, typically an alcohol-gasoline blend such as E85 or M85 (85 percent ethanol or methanol, respectively, with 15 percent gasoline).

We have not provided separate listings for bi- and flex-fuel vehicles since they are not generally certified to different emissions standards than their gasoline-only counterparts. A bi-fuel CNG vehicle will, however, generally be cleaner when running on CNG than when running on gasoline. FFVs may be cleaner when running on their alternative fuel, but in practice are mostly run on gasoline. We denote the availability of bi- and flex-fuel vehicles with footnotes in the tables that follow.

Green Ratings for 1999 Vehicles

SAMPLE LISTING AND *GREEN GUIDE TO CARS AND TRUCKS* KEY

Emission Standard.

Federal certification:

Tier 1	Current national standard.
Tier 1-D	National standard for diesels.
HDT	Heavy Duty Truck

California certification (shaded):

TLEV	Transitional Low Emission Vehicle
LEV	Low Emission Vehicle
ULEV	Ultra Low Emission Vehicle
SULEV	Super Ultra Low Emission Vehicle
ZEV	Zero Emission Vehicle

Fuel Economy, City and **Hwy.** City and highway fuel economy in miles per gallon (MPG), adjusted to better reflect real-world driving conditions according to an EPA formula.

Values for Compressed Natural Gas (CNG) vehicles are in equivalent-MPG. Values for electric vehicles are in miles per kilowatt-hour (mi/kWh).

Manufacturer and model.

	Emission Standard	Fuel Economy City	Hwy
MERCEDES ML320			
3.2L 6, auto [P]	LEV	17	21

Number of Cylinders.

Engine Displacement. Specified in liters (L).

Transmission. Manual or automatic transmission is indicated. For models with two or more automatic transmissions available, the number of speeds is given. A few vehicles have continuously variable automatic transmissions, indicated by CVT.

Fuel Type. All vehicles use regular gasoline unless indicated by:

[P]	Premium Gasoline
[D]	Diesel
[CNG]	Compressed Natural Gas
Electric	Electric Vehicle
FFV	Flexible Fuel Vehicle

Note: Automakers use a variety of trimline designations, such as "DX," "LX," "SS," etc., on their vehicles, but their meaning is not uniform across makes and models. Therefore, the *Green Guide to Cars and Trucks* relies on engine and transmission specifications to distinguish among models.

Class Ranking. This rating compares vehicles *within the same class.*

- ✔ Superior
- ▲ Above Average
- ○ Average
- ▽ Below Average
- ✖ Inferior

Note: No vehicle can get a ✔ rating if its EDX is worse than the average for all 1999 vehicles, even if it is the best in its class.

Fuel Cost/yr. Estimated annual fuel costs based on 15,000 miles of driving and national average fuel prices.

Health Cost/yr. Estimated annual health costs from pollution based on damage cost estimates and emissions of CO, HC, NO_x and PM. This value represents the annual cost imposed on society due to illnesses and premature deaths in proportion to the amount of pollution produced by the vehicle.

Fuel Cost/yr	Health Cost/yr	GHG tons/yr	EDX ¢/mile	Green Score	Class Ranking
$1,080	$230	15	3.21	**16**	○

GHG tons/yr. Tons of CO_2 and other greenhouse gasses (GHG) emitted per year. Other GHGs are weighted according to their global warming impact. For example, a pound of methane has the same impact as about 22 pounds of CO_2.

EDX ¢/mile. The Environmental Damage Index represents a vehicle's overall environmental impact. The EDX is expressed as the environmental cost to society from each mile the vehicle travels (in cents per mile) accounting for the health costs and the risks associated with global warming.

Green Score. The EDX is converted to a 0-100 Green Score scale, where 100 represents the ideal of a pollution-free vehicle. The Green Score can be used to compare vehicles across different classes.

	Emission Standard	Fuel Economy City	Fuel Economy Hwy	Fuel Cost/yr	Health Cost/yr	GHG tons/yr	EDX ¢/mi	Green Score	Class Ranking
TWO SEATERS									
ACURA NSX									
3.0L 6, auto [P]	Tier 1	18	24	$980	$200	13	2.70	**20**	▽
3.2L 6, manual [P]	Tier 1	17	24	$980	$200	13	2.72	**20**	▽
BMW M COUPE									
3.1L 6, manual	Tier 1	19	26	$790	$190	12	2.56	**21**	○
BMW M ROADSTER									
3.1L 6, manual	Tier 1	19	26	$790	$190	12	2.56	**21**	○
BMW Z3 COUPE									
2.8L 6, manual	LEV	19	26	$820	$170	12	2.37	**23**	○
2.8L 6, auto	LEV	19	26	$790	$170	12	2.42	**23**	○
2.8L 6, manual	Tier 1	19	26	$820	$190	12	2.53	**22**	○
2.8L 6, auto	Tier 1	19	26	$790	$190	12	2.58	**21**	○
BMW Z3 ROADSTER									
2.5L 6, manual	LEV	20	27	$790	$160	12	2.31	**24**	▲
2.5L 6, auto	LEV	19	26	$820	$170	12	2.37	**23**	○
2.8L 6, auto	LEV	19	26	$820	$170	12	2.37	**23**	○
2.8L 6, manual	LEV	19	26	$820	$170	12	2.37	**23**	○
2.5L 6, manual	Tier 1	20	27	$790	$190	12	2.47	**22**	○
2.5L 6, auto	Tier 1	19	26	$820	$190	12	2.53	**22**	○
2.8L 6, auto	Tier 1	19	26	$820	$190	12	2.53	**22**	○
2.8L 6, manual	Tier 1	19	26	$820	$190	12	2.53	**22**	○
CHEVROLET CORVETTE									
5.7L 8, manual [P]	Tier 1	18	28	$890	$190	12	2.57	**21**	○
5.7L 8, auto [P]	Tier 1	17	25	$1,030	$200	13	2.75	**19**	▽
FERRARI 550 MARANE									
5.5L 12, manual [P]	Tier 1	9	14	$1,770	$270	23	4.17	**9**	✖
FERRARI F355/355 F1									
3.5L 8, manual [P]	Tier 1	10	15	$1,620	$250	21	3.82	**11**	✖
3.5L 8, auto [P]	Tier 1	10	14	$1,770	$260	21	3.93	**11**	✖
GM EV-1									
Electric[1]	ZEV	3.3	4.0	$410	$60	4	0.85	**57**	✔
LAMBORGHINI DB132/144 DIABLO									
5.7L 12, manual [P]	Tier 1	10	13	$1,770	$260	22	4.03	**10**	✖

[1] *Electric vehicle fuel economy is given in mi/kWh.*

	Emission Standard	Fuel Economy City	Fuel Economy Hwy	Fuel Cost/yr	Health Cost/yr	GHG tons/yr	EDX ¢/mi	Green Score	Class Ranking
TWO SEATERS (cont.)									
MAZDA MX-5 MIATA									
1.8L 4, manual	TLEV	25	29	$640	$170	10	2.13	**27**	▲
1.8L 4, manual	Tier 1	25	29	$640	$170	10	2.16	**26**	▲
1.8L 4, auto	TLEV	23	28	$690	$170	11	2.23	**25**	▲
1.8L 4, auto	Tier 1	23	28	$690	$170	11	2.26	**25**	▲
MERCEDES SL500									
5.0L 8, auto [P]	LEV	16	23	$1,030	$190	14	2.77	**19**	▽
5.0L 8, auto [P]	Tier 1	16	23	$1,030	$220	14	2.93	**17**	▽
MERCEDES SL600									
6.0L 2, auto [P]	Tier 1	13	19	$1,300	$230	17	3.34	**14**	✖
MERCEDES SLK230 KOMPRESSOR									
2.3L 4, auto [P]	TLEV	22	30	$780	$180	11	2.36	**24**	○
2.3L 4, auto [P]	Tier 1	22	30	$780	$190	11	2.39	**23**	○
PLYMOUTH PROWLER									
3.5L 6, auto [P]	Tier 1	17	23	$1,030	$200	13	2.70	**20**	▽
PORSCHE BOXSTER									
2.5L 6, manual [P]	Tier 1	19	26	$890	$190	12	2.51	**22**	○
2.5L 6, auto [P]	Tier 1	17	24	$1,030	$200	13	2.71	**20**	▽
SUBCOMPACT CARS									
ACURA 2.3CL/3.0CL									
2.3L 4, manual	Tier 1	24	31	$640	$180	10	2.27	**25**	○
2.3L 4, auto	Tier 1	22	30	$690	$180	11	2.37	**24**	○
3.0L 6, auto	Tier 1	20	28	$750	$190	12	2.51	**22**	▽
ACURA INTEGRA									
1.8L 4, manual [P]	Tier 1	25	31	$720	$170	10	2.17	**26**	○
1.8L 4, manual	Tier 1	25	32	$620	$170	10	2.19	**26**	○
1.8L 4, auto	Tier 1	24	32	$640	$170	10	2.21	**26**	○
BENTLEY AZURE									
6.8L 8, auto [P]	Tier 1	11	16	$1,500	$270	20	3.90	**11**	✖
BENTLEY CONTINENTAL									
6.8L 8, auto [P]	Tier 1	11	16	$1,500	$260	20	3.83	**11**	✖

See pages 32-33 for explanation of data and symbols.

	Emission Standard	Fuel Economy City	Fuel Economy Hwy	Fuel Cost/yr	Health Cost/yr	GHG tons/yr	EDX ¢/mi	Green Score	Class Ranking
SUBCOMPACT CARS (cont.)									
BMW 323I CONVERTIBLE									
2.5L 6, manual	Tier 1	21	31	$720	$190	11	2.42	**23**	▽
2.5L 6, auto	Tier 1	19	27	$790	$190	12	2.55	**21**	▽
BMW 323IS									
2.5L 6, manual	Tier 1	21	31	$720	$190	11	2.42	**23**	▽
2.5L 6, auto	Tier 1	19	27	$790	$190	12	2.55	**21**	▽
BMW 328I CONVERTIBLE									
2.8L 6, manual	TLEV	20	29	$750	$190	12	2.44	**23**	▽
2.8L 6, manual	Tier 1	20	29	$750	$190	12	2.47	**22**	▽
2.8L 6, auto	TLEV	20	27	$750	$190	12	2.49	**22**	▽
2.8L 6, auto	Tier 1	20	27	$750	$190	12	2.52	**22**	▽
BMW 328IS									
2.8L 6, manual	TLEV	20	29	$750	$190	12	2.44	**23**	▽
2.8L 6, manual	Tier 1	20	29	$750	$190	12	2.47	**22**	▽
2.8L 6, auto	TLEV	20	27	$750	$190	12	2.49	**22**	▽
2.8L 6, auto	Tier 1	20	27	$750	$190	12	2.52	**22**	▽
BMW M3									
3.1L 6, manual	TLEV	19	26	$790	$190	12	2.53	**22**	▽
3.1L 6, manual	Tier 1	19	26	$790	$190	12	2.56	**21**	▽
BMW M3 CONVERTIBLE									
3.1L 6, manual	Tier 1	19	26	$790	$190	12	2.56	**21**	▽
3.1L 6, auto	Tier 1	17	25	$860	$200	13	2.75	**19**	✖
CHEVROLET CAMARO									
3.8L 6, manual	TLEV	19	30	$750	$190	12	2.46	**22**	▽
3.8L 6, manual	Tier 1	19	30	$750	$190	12	2.49	**22**	▽
3.8L 6, auto	TLEV	19	29	$790	$190	12	2.50	**22**	▽
3.8L 6, auto	Tier 1	19	29	$790	$190	12	2.53	**22**	▽
5.7L 8, manual [P]	TLEV	19	28	$890	$190	12	2.53	**22**	▽
3.8L 6, manual	Tier 1	19	30	$750	$200	12	2.54	**21**	▽
3.8L 6, auto	TLEV	19	29	$790	$190	12	2.55	**21**	▽
5.7L 8, manual [P]	Tier 1	19	28	$890	$190	12	2.56	**21**	▽
5.7L 8, manual [P]	Tier 1	19	28	$890	$200	12	2.62	**21**	✖
5.7L 8, auto [P]	TLEV	17	24	$980	$200	13	2.71	**20**	✖
5.7L 8, auto [P]	Tier 1	17	24	$980	$200	13	2.74	**19**	✖
5.7L 8, auto [P]	TLEV	17	24	$980	$200	14	2.76	**19**	✖

See pages 32-33 for explanation of data and symbols.

	Emission Standard	Fuel Economy City	Fuel Economy Hwy	Fuel Cost/yr	Health Cost/yr	GHG tons/yr	EDX ¢/mi	Green Score	Class Ranking
SUBCOMPACT CARS (cont.)									
CHEVROLET CAVALIER [2]									
2.2L 4, manual	LEV	24	34	$620	$150	10	2.04	**28**	○
2.4L 4, manual	LEV	23	33	$660	$150	10	2.09	**27**	○
2.2L 4, auto 4 sp	LEV	23	31	$660	$150	10	2.10	**27**	○
2.4L 4, auto 4 sp	LEV	22	30	$690	$160	11	2.15	**27**	○
2.2L 4, auto 3 sp	LEV	23	29	$690	$160	11	2.15	**26**	○
2.2L 4, manual	TLEV	24	34	$620	$170	10	2.17	**26**	○
2.2L 4, manual	Tier 1	24	34	$620	$170	10	2.19	**26**	○
2.4L 4, manual	TLEV	23	33	$660	$170	10	2.22	**25**	○
2.2L 4, auto 4 sp	TLEV	23	31	$660	$170	10	2.23	**25**	○
2.4L 4, manual	Tier 1	23	33	$660	$180	10	2.25	**25**	○
2.2L 4, auto 4 sp	Tier 1	23	31	$660	$180	10	2.26	**25**	○
2.4L 4, auto 4 sp	TLEV	22	30	$690	$180	11	2.28	**25**	○
2.2L 4, auto 3 sp	TLEV	23	29	$690	$180	11	2.29	**25**	○
2.4L 4, auto 4 sp	Tier 1	22	30	$690	$180	11	2.31	**24**	○
2.2L 4, auto 3 sp	Tier 1	23	29	$690	$180	11	2.31	**24**	○
CHEVROLET METRO									
1.0L 3, manual	LEV	41	47	$400	$120	7	1.50	**38**	✔
1.3L 4, manual	LEV	39	43	$420	$130	7	1.57	**37**	✔
1.0L 3, manual	TLEV	41	47	$400	$140	7	1.63	**36**	✔
1.0L 3, manual	Tier 1	41	47	$400	$140	7	1.65	**35**	✔
1.3L 4, manual	TLEV	39	43	$420	$150	7	1.70	**34**	✔
1.3L 4, manual	Tier 1	39	43	$420	$150	7	1.73	**34**	✔
1.3L 4, auto	LEV	30	34	$540	$140	9	1.79	**33**	▲
1.3L 4, auto	TLEV	30	34	$540	$150	9	1.92	**30**	▲
1.3L 4, auto	Tier 1	30	34	$540	$160	9	1.94	**30**	▲
FERRARI 456 MGT/MGTA									
5.5L 12, manual [P]	Tier 1	10	16	$1,620	$260	21	3.94	**11**	✖
5.5L 12, auto [P]	Tier 1	10	15	$1,770	$270	22	4.04	**10**	✖
FORD ESCORT ZX2									
2.0L 4, manual	Tier 1	25	33	$620	$170	10	2.13	**27**	○
2.0L 4, auto	Tier 1	25	33	$620	$170	10	2.15	**27**	○
FORD MUSTANG									
3.8L 6, manual	Tier 1	20	29	$750	$190	12	2.48	**22**	▽
3.8L 6, auto	Tier 1	20	27	$790	$190	12	2.52	**22**	▽
4.6L 8, auto	Tier 1	17	23	$860	$200	13	2.73	**19**	✖
4.6L 8, manual	Tier 1	17	24	$860	$200	13	2.74	**19**	✖

[2] *A compressed natural gas (CNG)–gasoline bi-fuel version of this vehicle is also available.*

	Emission Standard	Fuel Economy City	Fuel Economy Hwy	Fuel Cost/yr	Health Cost/yr	GHG tons/yr	EDX ¢/mi	Green Score	Class Ranking
SUBCOMPACT CARS (cont.)									
HONDA CIVIC									
1.6L 4, manual	LEV	32	37	$510	$140	8	1.76	**33**	▲
1.6L 4, manual	LEV	29	35	$540	$140	9	1.86	**31**	▲
1.6L 4, auto	LEV	28	35	$560	$140	9	1.88	**31**	▲
1.6L 4, manual	Tier 1	32	37	$510	$160	8	1.91	**30**	▲
1.6L 4, manual [P]	LEV	26	31	$700	$150	10	1.99	**29**	▲
1.6L 4, manual	Tier 1	29	35	$540	$160	9	2.01	**29**	○
1.6L 4, auto	Tier 1	28	35	$560	$170	9	2.04	**28**	○
1.6L 4, auto	Tier 1	28	35	$570	$170	9	2.06	**28**	○
1.6L 4, manual [P]	Tier 1	26	31	$700	$170	10	2.15	**27**	○
HONDA CIVIC GX									
1.6L 4, auto [CNG][3]	ULEV	28	34	$330	$70	7	1.19	**46**	✔
HONDA CIVIC HX									
1.6L 4, manual	Tier 1	35	43	$450	$150	8	1.81	**32**	▲
1.6L 4, CVT	Tier 1	34	38	$480	$160	8	1.90	**31**	▲
HONDA PRELUDE									
2.2L 4, manual [P]	Tier 1	22	27	$810	$190	11	2.42	**23**	▽
2.2L 4, auto [P]	Tier 1	21	26	$850	$190	12	2.47	**22**	▽
HYUNDAI ACCENT									
1.5L 4, manual	TLEV	28	37	$540	$160	9	1.96	**29**	▲
1.5L 4, manual	Tier 1	28	37	$540	$160	9	1.99	**29**	▲
1.5L 4, auto	TLEV	26	36	$570	$160	9	2.03	**28**	○
1.5L 4, auto	Tier 1	26	36	$570	$170	9	2.06	**28**	○
HYUNDAI TIBURON									
2.0L 4, manual	LEV	22	31	$660	$160	11	2.12	**27**	○
2.0L 4, auto	LEV	22	29	$690	$160	11	2.16	**26**	○
2.0L 4, manual	Tier 1	22	31	$660	$180	11	2.28	**25**	○
2.0L 4, auto	Tier 1	22	29	$690	$180	11	2.32	**24**	○
JAGUAR XK8									
4.0L 8, auto [P]	Tier 1	17	25	$980	$210	14	2.78	**19**	✖
LEXUS SC 300/SC 400									
3.0L 6, auto 4 sp [P]	TLEV	19	24	$930	$200	13	2.66	**20**	✖
4.0L 8, auto 5 sp [P]	TLEV	18	25	$930	$200	13	2.66	**20**	✖
3.0L 6, auto 4 sp [P]	Tier 1	19	24	$930	$200	13	2.69	**20**	✖
4.0L 8, auto 5 sp [P]	Tier 1	18	25	$930	$200	13	2.69	**20**	✖

[3] *CNG fuel economy is given as gasoline-equivalent MPG.*

	Emission Standard	Fuel Economy City	Fuel Economy Hwy	Fuel Cost/yr	Health Cost/yr	GHG tons/yr	EDX ¢/mi	Green Score	Class Ranking
SUBCOMPACT CARS (cont.)									
MERCEDES CLK320									
3.2L 6, auto [P]	LEV	21	29	$810	$170	12	2.30	**24**	○
3.2L 6, auto [P]	Tier 1	21	29	$810	$190	12	2.45	**22**	▽
MERCEDES CLK320 CABRIOLET									
3.2L 6, auto [P]	Tier 1	19	28	$890	$200	12	2.58	**21**	▽
MERCEDES CLK430									
4.3L 8, auto [P]	LEV	18	25	$930	$170	13	2.48	**22**	▽
4.3L 8, auto [P]	Tier 1	18	25	$930	$200	13	2.64	**20**	✖
MITSUBISHI 3000GT									
3.0L 6, manual	Tier 1	19	25	$820	$200	12	2.60	**21**	▽
3.0L 6, manual	Tier 1	19	24	$820	$200	13	2.62	**21**	✖
3.0L 6, auto	Tier 1	18	23	$860	$200	13	2.67	**20**	✖
3.0L 6, auto	Tier 1	18	24	$860	$200	13	2.67	**20**	✖
3.0L 6, manual [P]	Tier 1	18	24	$980	$200	13	2.74	**19**	✖
MITSUBISHI ECLIPSE									
2.0L 4, manual	LEV	22	33	$660	$150	10	2.11	**27**	○
2.0L 4, manual [P]	LEV	23	31	$750	$160	11	2.15	**26**	○
2.3L 4, manual	LEV	22	30	$690	$160	11	2.17	**26**	○
2.0L 4, auto	LEV	21	30	$720	$160	11	2.21	**26**	○
2.3L 4, auto	LEV	21	28	$750	$160	11	2.25	**25**	○
2.0L 4, manual	Tier 1	22	33	$660	$180	10	2.27	**25**	○
2.0L 4, manual [P]	LEV	21	28	$810	$160	11	2.28	**25**	○
2.0L 4, manual [P]	Tier 1	23	31	$750	$180	11	2.31	**24**	○
2.3L 4, manual	Tier 1	22	30	$690	$180	11	2.33	**24**	○
2.0L 4, auto [P]	LEV	20	27	$850	$170	12	2.34	**24**	○
2.0L 4, auto	Tier 1	21	30	$720	$180	11	2.37	**23**	○
2.3L 4, auto	Tier 1	21	28	$750	$180	11	2.41	**23**	▽
2.0L 4, auto [P]	LEV	19	25	$930	$170	12	2.43	**23**	▽
2.0L 4, manual [P]	Tier 1	21	28	$810	$190	11	2.43	**23**	▽
2.0L 4, auto [P]	Tier 1	20	27	$850	$190	12	2.50	**22**	▽
2.0L 4, auto [P]	Tier 1	19	25	$930	$200	12	2.59	**21**	▽
MITSUBISHI MIRAGE									
1.5L 4, manual	LEV	33	40	$480	$130	8	1.71	**34**	✔
1.8L 4, manual	LEV	28	36	$560	$140	9	1.83	**32**	▲
1.5L 4, auto	LEV	28	36	$560	$140	9	1.84	**32**	▲
1.5L 4, manual	Tier 1	33	40	$480	$160	8	1.87	**31**	▲
1.8L 4, auto	LEV	26	33	$600	$140	9	1.93	**30**	▲

See pages 32-33 for explanation of data and symbols.

	Emission Standard	Fuel Economy City	Fuel Economy Hwy	Fuel Cost/yr	Health Cost/yr	GHG tons/yr	EDX ¢/mi	Green Score	Class Ranking
SUBCOMPACT CARS (cont.)									
MITSUBISHI MIRAGE (cont.)									
1.8L 4, manual	Tier 1	28	36	$560	$160	9	1.99	**29**	▲
1.5L 4, auto	Tier 1	28	36	$560	$160	9	2.00	**29**	▲
1.8L 4, auto	Tier 1	26	33	$600	$170	9	2.09	**27**	○
NISSAN SENTRA/200SX									
1.6L 4, manual	LEV	29	39	$520	$140	8	1.79	**32**	▲
1.6L 4, auto	LEV	27	36	$570	$140	9	1.90	**30**	▲
1.6L 4, manual	Tier 1	29	39	$520	$160	8	1.95	**30**	▲
1.6L 4, auto	Tier 1	27	36	$570	$170	9	2.06	**28**	○
2.0L 4, manual	LEV	23	31	$660	$150	10	2.10	**27**	○
2.0L 4, auto	LEV	23	30	$660	$160	11	2.13	**27**	○
2.0L 4, manual	Tier 1	23	31	$660	$180	10	2.25	**25**	○
2.0L 4, auto	Tier 1	23	30	$660	$180	11	2.29	**25**	○
PONTIAC FIREBIRD/TRANS AM									
3.8L 6, manual	TLEV	19	30	$750	$190	12	2.46	**22**	▽
3.8L 6, manual	Tier 1	19	30	$750	$190	12	2.49	**22**	▽
3.8L 6, auto	TLEV	19	29	$790	$190	12	2.49	**22**	▽
3.8L 6, auto	Tier 1	19	29	$790	$190	12	2.52	**22**	▽
5.7L 8, manual [P]	TLEV	19	28	$890	$190	12	2.57	**21**	▽
5.7L 8, manual [P]	Tier 1	19	28	$890	$200	12	2.60	**21**	▽
5.7L 8, auto [P]	TLEV	18	24	$980	$200	13	2.70	**20**	✖
5.7L 8, auto [P]	Tier 1	18	24	$980	$200	13	2.73	**19**	✖
PONTIAC SUNFIRE									
2.2L 4, manual	LEV	24	34	$620	$150	10	2.04	**28**	○
2.4L 4, manual	LEV	23	33	$660	$150	10	2.09	**27**	○
2.2L 4, auto 4 sp	LEV	23	31	$660	$150	10	2.10	**27**	○
2.4L 4, auto 4 sp	LEV	22	30	$690	$160	11	2.15	**27**	○
2.2L 4, auto 3 sp	LEV	23	29	$690	$160	11	2.15	**26**	○
2.2L 4, manual	TLEV	24	34	$620	$170	10	2.17	**26**	○
2.2L 4, manual	Tier 1	24	34	$620	$170	10	2.19	**26**	○
2.4L 4, manual	TLEV	23	33	$660	$170	10	2.22	**25**	○
2.2L 4, auto 4 sp	TLEV	23	31	$660	$170	10	2.23	**25**	○
2.4L 4, manual	Tier 1	23	33	$660	$180	10	2.25	**25**	○
2.2L 4, auto 4 sp	Tier 1	23	31	$660	$180	10	2.26	**25**	○
2.4L 4, auto 4 sp	TLEV	22	30	$690	$180	11	2.28	**25**	○
2.2L 4, auto 3 sp	TLEV	23	29	$690	$180	11	2.29	**25**	○
2.4L 4, auto 4 sp	Tier 1	22	30	$690	$180	11	2.31	**24**	○

See pages 32-33 for explanation of data and symbols.

	Emission Standard	Fuel Economy City	Fuel Economy Hwy	Fuel Cost/yr	Health Cost/yr	GHG tons/yr	EDX ¢/mi	Green Score	Class Ranking
SUBCOMPACT CARS (cont.)									
PONTIAC SUNFIRE (cont.)									
2.2L 4, auto 3 sp	Tier 1	23	29	$690	$180	11	2.31	**24**	○
2.4L 4, auto 4 sp	Tier 1	22	30	$690	$180	11	2.36	**24**	○
PORSCHE 911 CARRERA									
3.4L 6, manual [P]	Tier 1	17	25	$980	$200	13	2.71	**20**	✖
3.4L 6, auto [P]	Tier 1	16	25	$1,030	$200	13	2.75	**19**	✖
SAAB 9-3 CONVERTIBLE									
2.0L 4, manual	Tier 1	20	27	$750	$190	12	2.49	**22**	▽
2.0L 4, manual [P]	Tier 1	19	27	$890	$190	12	2.54	**21**	▽
2.0L 4, auto	Tier 1	19	25	$820	$200	13	2.62	**21**	✖
SATURN SC									
1.9L 4, manual	LEV	29	40	$520	$140	8	1.79	**33**	▲
1.9L 4, auto	LEV	27	37	$570	$140	9	1.88	**31**	▲
1.9L 4, manual	LEV	27	38	$560	$140	9	1.89	**31**	▲
1.9L 4, manual	TLEV	29	40	$520	$160	8	1.92	**30**	▲
1.9L 4, manual	Tier 1	29	40	$520	$160	8	1.94	**30**	▲
1.9L 4, auto	LEV	25	35	$600	$150	9	1.96	**30**	▲
1.9L 4, auto	TLEV	27	37	$570	$160	9	2.01	**29**	○
1.9L 4, manual	TLEV	27	38	$560	$160	9	2.02	**29**	○
1.9L 4, auto	Tier 1	27	37	$570	$160	9	2.03	**28**	○
1.9L 4, manual	Tier 1	27	38	$560	$170	9	2.04	**28**	○
1.9L 4, auto	TLEV	25	35	$600	$170	10	2.09	**27**	○
1.9L 4, auto	Tier 1	25	35	$600	$170	10	2.12	**27**	○
SUBARU IMPREZA									
2.2L 4, auto	LEV	23	29	$690	$160	11	2.15	**27**	○
2.2L 4, manual	LEV	22	29	$690	$160	11	2.15	**26**	○
2.5L 4, auto	LEV	22	28	$690	$160	11	2.18	**26**	○
2.5L 4, manual	LEV	22	29	$720	$160	11	2.19	**26**	○
2.2L 4, auto	TLEV	23	29	$690	$180	11	2.28	**25**	○
2.2L 4, manual	TLEV	22	29	$690	$180	11	2.29	**25**	○
2.5L 4, auto	TLEV	22	28	$690	$180	11	2.31	**24**	○
2.2L 4, auto	Tier 1	23	29	$690	$180	11	2.31	**24**	○
2.2L 4, manual	Tier 1	22	29	$690	$180	11	2.31	**24**	○
2.5L 4, manual	TLEV	22	29	$720	$180	11	2.32	**24**	○
2.5L 4, auto	Tier 1	22	28	$690	$180	11	2.33	**24**	○
2.5L 4, manual	Tier 1	22	29	$720	$180	11	2.34	**24**	○

See pages 32-33 for explanation of data and symbols.

SUBCOMPACT CARS (cont.)

	Emission Standard	Fuel Economy City	Fuel Economy Hwy	Fuel Cost/yr	Health Cost/yr	GHG tons/yr	EDX ¢/mi	Green Score	Class Ranking
SUZUKI ESTEEM									
1.6L 4, manual	TLEV	30	37	$520	$160	8	1.92	**30**	▲
1.6L 4, manual	Tier 1	30	37	$520	$160	8	1.94	**30**	▲
2.0L 4, manual	TLEV	28	35	$560	$160	9	1.98	**29**	▲
2.0L 4, manual	Tier 1	28	35	$560	$160	9	2.01	**29**	○
1.6L 4, auto	TLEV	27	34	$570	$160	9	2.02	**29**	○
1.6L 4, auto	Tier 1	27	34	$570	$160	9	2.05	**28**	○
SUZUKI SWIFT									
1.3L 4, manual	LEV	39	43	$420	$130	7	1.57	**37**	✔
1.3L 4, manual	Tier 1	39	43	$420	$150	7	1.73	**34**	✔
1.3L 4, auto	LEV	30	34	$540	$140	9	1.79	**33**	▲
1.3L 4, auto	Tier 1	30	34	$540	$160	9	1.94	**30**	▲
TOYOTA CELICA									
2.2L 4, auto	Tier 1	22	28	$690	$180	11	2.33	**24**	○
2.2L 4, manual	Tier 1	22	28	$720	$180	11	2.35	**24**	○
TOYOTA CELICA CONVERTIBLE									
2.2L 4, auto	Tier 1	23	30	$660	$180	11	2.28	**25**	○
2.2L 4, manual	Tier 1	22	28	$720	$180	11	2.35	**24**	○
TOYOTA PASEO									
1.5L 4, manual	Tier 1	33	41	$480	$160	8	1.86	**31**	▲
1.5L 4, auto	Tier 1	27	32	$600	$170	9	2.07	**28**	○
TOYOTA PASEO CONVERTIBLE									
1.5L 4, manual	Tier 1	33	41	$480	$160	8	1.86	**31**	▲
1.5L 4, auto	Tier 1	27	32	$600	$170	9	2.07	**28**	○
TOYOTA TERCEL									
1.5L 4, manual	Tier 1	32	40	$490	$150	8	1.85	**31**	▲
1.5L 4, manual	Tier 1	32	40	$490	$160	8	1.87	**31**	▲
1.5L 4, auto 4 sp	Tier 1	30	37	$520	$160	9	1.95	**30**	▲
1.5L 4, auto 3 sp	Tier 1	29	33	$560	$160	9	2.00	**29**	○
VOLKSWAGEN CABRIO									
2.0L 4, manual	TLEV	24	31	$640	$170	10	2.21	**26**	○
2.0L 4, manual	Tier 1	24	31	$640	$180	10	2.24	**25**	○
2.0L 4, auto	TLEV	22	28	$720	$180	11	2.34	**24**	○
2.0L 4, auto	Tier 1	22	28	$720	$180	11	2.36	**24**	○

See pages 32-33 for explanation of data and symbols.

	Emission Standard	Fuel Economy City	Hwy	Fuel Cost/yr	Health Cost/yr	GHG tons/yr	EDX ¢/mi	Green Score	Class Ranking
SUBCOMPACT CARS (cont.)									
VOLKSWAGEN NEW BEETLE									
2.0L 4, manual	Tier 1	24	31	$640	$180	10	2.23	**25**	○
2.0L 4, auto	Tier 1	22	28	$720	$180	11	2.34	**24**	○
1.9L 4, manual [D]	Tier 1-D	42	49	$370	$260	6	2.37	**24**	○
1.9L 4, auto [D]	Tier 1-D	34	45	$430	$270	7	2.49	**22**	▽
VOLVO C70 CONVERTIBLE									
2.3L 5, manual [P]	Tier 1	20	27	$890	$200	12	2.58	**21**	▽
2.3L 5, auto [P]	Tier 1	19	26	$890	$200	12	2.60	**21**	▽
2.4L 5, auto [P]	Tier 1	19	27	$890	$200	12	2.62	**21**	✖
COMPACT CARS									
AUDI A4									
1.8L 4, manual [P]	TLEV	23	32	$750	$170	10	2.21	**26**	○
1.8L 4, manual [P]	Tier 1	23	32	$750	$180	10	2.24	**25**	○
1.8L 4, auto [P]	TLEV	21	31	$810	$180	11	2.39	**23**	○
1.8L 4, auto [P]	Tier 1	21	31	$810	$190	11	2.41	**23**	▽
2.8L 6, manual [P]	TLEV	20	29	$850	$190	12	2.46	**22**	▽
2.8L 6, manual [P]	Tier 1	20	29	$850	$190	12	2.48	**22**	▽
2.8L 6, auto [P]	TLEV	18	29	$890	$190	12	2.55	**21**	✖
2.8L 6, auto [P]	Tier 1	18	29	$890	$190	12	2.58	**21**	✖
AUDI A4 QUATTRO									
1.8L 4, manual [P]	TLEV	21	29	$810	$180	11	2.40	**23**	○
1.8L 4, manual [P]	Tier 1	21	29	$810	$190	11	2.42	**23**	▽
2.8L 6, manual [P]	TLEV	19	27	$890	$190	12	2.52	**22**	✖
2.8L 6, manual [P]	Tier 1	19	27	$890	$190	12	2.55	**21**	✖
1.8L 4, auto 5 sp [P]	TLEV	18	27	$930	$190	12	2.56	**21**	✖
1.8L 4, auto 5 sp [P]	Tier 1	18	27	$930	$190	12	2.59	**21**	✖
2.8L 6, auto 5 sp [P]	TLEV	17	27	$930	$190	13	2.61	**21**	✖
2.8L 6, auto 5 sp [P]	Tier 1	17	27	$930	$200	13	2.63	**20**	✖
6.8L 8, auto 4 sp [P]	Tier 1	11	16	$1,500	$260	20	3.83	**11**	✖
BMW 318TI									
1.9L 4, manual	Tier 1	23	32	$640	$180	10	2.24	**25**	○
1.9L 4, auto	Tier 1	23	31	$660	$180	10	2.27	**25**	○

See pages 32-33 for explanation of data and symbols.

	Emission Standard	Fuel Economy City	Hwy	Fuel Cost/yr	Health Cost/yr	GHG tons/yr	EDX ¢/mi	Green Score	Class Ranking
COMPACT CARS (cont.)									
BMW 323I									
2.5L 6, manual	Tier 1	20	29	$750	$190	12	2.46	**22**	▽
2.5L 6, auto	Tier 1	19	28	$790	$190	12	2.53	**22**	✖
BMW 328I									
2.8L 6, manual	TLEV	20	29	$720	$180	12	2.43	**23**	▽
2.8L 6, manual	Tier 1	20	29	$720	$190	12	2.45	**22**	▽
2.8L 6, auto	TLEV	19	27	$790	$190	12	2.52	**22**	✖
2.8L 6, auto	Tier 1	19	27	$790	$190	12	2.54	**21**	✖
BMW 528I									
2.8L 6, manual	LEV	20	29	$720	$170	12	2.30	**24**	○
2.8L 6, manual	Tier 1	20	29	$720	$190	12	2.45	**22**	▽
2.8L 6, auto	LEV	18	26	$820	$180	13	2.52	**22**	✖
2.8L 6, auto	Tier 1	18	26	$820	$200	13	2.68	**20**	✖
BMW 540I									
4.4L 8, auto	Tier 1	18	24	$860	$200	13	2.72	**19**	✖
4.4L 8, manual	Tier 1	15	23	$960	$210	15	2.93	**17**	✖
4.4L 8, auto	Tier 1	15	21	$960	$220	15	2.98	**17**	✖
CHEVROLET PRIZM									
1.8L 4, manual	TLEV	31	37	$520	$160	8	1.93	**30**	▲
1.8L 4, manual	Tier 1	31	37	$520	$160	8	1.95	**30**	▲
1.8L 4, auto 4 sp	TLEV	28	36	$560	$160	9	2.00	**29**	▲
1.8L 4, auto 4 sp	Tier 1	28	36	$560	$160	9	2.02	**28**	▲
1.8L 4, auto 3 sp	TLEV	28	33	$570	$160	9	2.04	**28**	▲
1.8L 4, auto 3 sp	Tier 1	28	33	$570	$170	9	2.07	**28**	▲
CHRYSLER SEBRING									
2.0L 4, manual	Tier 1	22	31	$690	$180	11	2.36	**24**	○
2.0L 4, auto	Tier 1	21	30	$720	$190	11	2.42	**23**	▽
2.5L 6, auto	Tier 1	19	27	$790	$190	12	2.54	**21**	✖
CHRYSLER SEBRING CONVERTIBLE									
2.5L 6, auto	LEV	19	27	$790	$170	12	2.40	**23**	▽
2.5L 6, auto	TLEV	19	27	$790	$190	12	2.53	**22**	✖
2.5L 6, auto	Tier 1	19	27	$790	$190	12	2.56	**21**	✖
DAEWOO LANOS									
1.5L 4, manual	TLEV	26	36	$570	$160	9	2.04	**28**	▲
1.6L 4, manual	TLEV	26	36	$570	$160	9	2.06	**28**	▲
1.5L 4, manual	Tier 1	26	36	$570	$170	9	2.07	**28**	▲

See pages 32-33 for explanation of data and symbols.

	Emission Standard	Fuel Economy City	Hwy	Fuel Cost/yr	Health Cost/yr	GHG tons/yr	EDX ¢/mi	Green Score	Class Ranking
COMPACT CARS (cont.)									
DAEWOO LANOS (cont.)									
1.6L 4, manual	Tier 1	26	36	$570	$170	9	2.09	**27**	▲
1.5L 4, auto	TLEV	24	37	$620	$170	10	2.12	**27**	○
1.5L 4, auto	Tier 1	24	37	$620	$170	10	2.14	**27**	○
1.6L 4, auto	TLEV	23	34	$640	$170	10	2.18	**26**	○
1.6L 4, auto	Tier 1	23	34	$640	$170	10	2.20	**26**	○
DAEWOO NUBIRA									
2.0L 4, manual	TLEV	22	31	$660	$170	10	2.25	**25**	○
2.0L 4, manual	Tier 1	22	31	$660	$180	11	2.27	**25**	○
2.0L 4, auto	TLEV	22	30	$690	$180	11	2.28	**25**	○
2.0L 4, auto	Tier 1	22	30	$690	$180	11	2.30	**24**	○
DODGE AVENGER									
2.0L 4, manual	TLEV	22	32	$690	$180	11	2.27	**25**	○
2.0L 4, manual	Tier 1	22	32	$690	$180	11	2.30	**24**	○
2.0L 4, auto	TLEV	21	30	$720	$180	11	2.39	**23**	○
2.0L 4, auto	Tier 1	21	30	$720	$190	11	2.42	**23**	▽
2.5L 6, auto	TLEV	19	27	$790	$190	12	2.51	**22**	✖
2.5L 6, auto	Tier 1	19	27	$790	$190	12	2.54	**21**	✖
DODGE NEON									
2.0L 4, manual	LEV	28	39	$540	$140	9	1.84	**32**	✔
2.0L 4, manual	Tier 1	28	39	$540	$160	9	2.00	**29**	▲
2.0L 4, auto	LEV	23	32	$640	$150	10	2.05	**28**	▲
2.0L 4, auto	Tier 1	23	32	$640	$170	10	2.21	**26**	○
FORD CONTOUR[4]									
2.0L 4, manual	Tier 1	24	34	$620	$170	10	2.18	**26**	○
2.0L 4, auto	Tier 1	23	31	$660	$180	11	2.28	**25**	○
2.5L 6, manual	Tier 1	20	28	$750	$190	12	2.44	**23**	▽
2.5L 6, auto	Tier 1	20	29	$750	$190	12	2.47	**22**	▽
FORD ESCORT									
2.0L 4, manual	LEV	28	37	$540	$140	9	1.86	**31**	✔
2.0L 4, auto	LEV	25	34	$600	$150	10	1.97	**29**	▲
2.0L 4, manual	Tier 1	28	37	$540	$160	9	2.02	**29**	▲
2.0L 4, auto	Tier 1	25	34	$600	$170	10	2.13	**27**	○
HONDA EV PLUS									
Electric[5]	ZEV	2.0	2.0	$750	$90	7	1.35	**42**	✔

[4] *A CNG-gasoline bi-fuel version of this vehicle is also available.*
[5] *Electric vehicle fuel economy is given in mi/kWh.*

	Emission Standard	Fuel Economy City	Fuel Economy Hwy	Fuel Cost/yr	Health Cost/yr	GHG tons/yr	EDX ¢/mi	Green Score	Class Ranking
COMPACT CARS (cont.)									
HYUNDAI ELANTRA SEDAN									
2.0L 4, manual	LEV	24	33	$640	$150	10	2.05	**28**	▲
2.0L 4, auto	LEV	22	31	$690	$160	11	2.16	**26**	○
2.0L 4, manual	Tier 1	24	33	$640	$170	10	2.20	**26**	○
2.0L 4, auto	Tier 1	22	31	$690	$180	11	2.32	**24**	○
INFINITI G20									
2.0L 4, manual	Tier 1	23	31	$660	$180	11	2.27	**25**	○
2.0L 4, auto	Tier 1	22	28	$720	$180	11	2.35	**24**	○
JAGUAR XJR									
4.0L 8, auto [P]	Tier 1	16	22	$1,080	$210	14	2.89	**18**	✖
KIA SEPHIA									
1.8L 4, manual	TLEV	24	31	$640	$170	10	2.16	**26**	○
1.8L 4, manual	Tier 1	24	31	$640	$170	10	2.19	**26**	○
1.8L 4, auto	TLEV	23	31	$660	$170	10	2.21	**26**	○
1.8L 4, auto	Tier 1	23	31	$660	$170	10	2.24	**25**	○
LEXUS ES 300									
3.0L 6, auto	TLEV	19	26	$820	$190	12	2.56	**21**	✖
3.0L 6, auto	Tier 1	19	26	$820	$200	12	2.59	**21**	✖
MAZDA MILLENIA									
2.3L 6, auto [P]	Tier 1	20	28	$890	$190	12	2.52	**22**	✖
2.5L 6, auto [P]	Tier 1	20	27	$890	$190	12	2.54	**21**	✖
MAZDA PROTEGE									
1.6L 4, manual	ULEV	29	34	$560	$140	9	1.82	**32**	✔
1.6L 4, manual	LEV	29	34	$560	$140	9	1.88	**31**	✔
1.6L 4, auto	ULEV	26	33	$600	$140	9	1.90	**31**	✔
1.6L 4, auto	LEV	26	33	$600	$150	9	1.96	**30**	▲
1.8L 4, manual	LEV	26	30	$620	$150	10	2.00	**29**	▲
1.6L 4, manual	TLEV	29	34	$560	$160	9	2.01	**29**	▲
1.6L 4, manual	Tier 1	29	34	$560	$170	9	2.04	**28**	▲
1.8L 4, auto	LEV	24	29	$660	$150	10	2.08	**27**	▲
1.6L 4, auto	TLEV	26	33	$600	$170	9	2.09	**27**	▲
1.6L 4, auto	Tier 1	26	33	$600	$170	10	2.11	**27**	○
1.8L 4, manual	TLEV	26	30	$620	$170	10	2.13	**27**	○
1.8L 4, manual	Tier 1	26	30	$620	$170	10	2.16	**26**	○
1.8L 4, auto	TLEV	24	29	$660	$170	10	2.22	**26**	○
1.8L 4, auto	Tier 1	24	29	$660	$180	10	2.24	**25**	○

See pages 32-33 for explanation of data and symbols.

	Emission Standard	Fuel Economy City	Fuel Economy Hwy	Fuel Cost/yr	Health Cost/yr	GHG tons/yr	EDX ¢/mi	Green Score	Class Ranking
COMPACT CARS (cont.)									
MERCEDES C230 KOMPRESSOR									
2.3L 4, auto [P]	TLEV	21	29	$810	$180	11	2.39	**23**	○
2.3L 4, auto [P]	Tier 1	21	29	$810	$190	11	2.41	**23**	▽
MERCEDES C280									
2.8L 6, auto [P]	LEV	21	27	$850	$170	12	2.31	**24**	○
2.8L 6, auto [P]	Tier 1	21	27	$850	$190	12	2.46	**22**	▽
MERCEDES C43 AMG									
4.3L 8, auto [P]	LEV	18	23	$980	$180	13	2.56	**21**	✖
4.3L 8, auto [P]	Tier 1	18	23	$980	$200	13	2.72	**20**	✖
MERCEDES CL500									
5.0L 8, auto [P]	TLEV	15	22	$1,080	$220	15	3.05	**16**	✖
5.0L 8, auto [P]	Tier 1	15	22	$1,080	$230	15	3.08	**16**	✖
MERCEDES CL600									
6.0L 2, auto [P]	Tier 1	13	19	$1,300	$240	17	3.41	**14**	✖
MERCURY COUGAR									
2.0L 4, manual	TLEV	24	34	$620	$170	10	2.15	**26**	○
2.0L 4, manual	Tier 1	24	34	$620	$170	10	2.18	**26**	○
2.0L 4, auto	TLEV	23	31	$660	$170	11	2.26	**25**	○
2.0L 4, auto	Tier 1	23	31	$660	$180	11	2.28	**25**	○
2.5L 6, auto	TLEV	20	29	$750	$190	12	2.44	**23**	▽
2.5L 6, auto	Tier 1	20	29	$750	$190	12	2.47	**22**	▽
2.5L 6, manual	TLEV	19	28	$790	$190	12	2.50	**22**	▽
2.5L 6, manual	Tier 1	19	28	$790	$190	12	2.53	**22**	✖
MERCURY MYSTIQUE									
2.0L 4, manual	TLEV	24	34	$620	$170	10	2.15	**26**	○
2.0L 4, manual	Tier 1	24	34	$620	$170	10	2.18	**26**	○
2.0L 4, auto	TLEV	23	31	$660	$170	11	2.26	**25**	○
2.0L 4, auto	Tier 1	23	31	$660	$180	11	2.28	**25**	○
2.5L 6, auto	TLEV	20	29	$750	$190	12	2.44	**23**	▽
2.5L 6, auto	Tier 1	20	29	$750	$190	12	2.47	**22**	▽
2.5L 6, manual	TLEV	19	28	$790	$190	12	2.50	**22**	▽
2.5L 6, manual	Tier 1	19	28	$790	$190	12	2.53	**22**	✖
MERCURY TRACER									
2.0L 4, manual	LEV	28	37	$540	$140	9	1.86	**31**	✔
2.0L 4, auto	LEV	25	34	$600	$150	10	1.97	**29**	▲

See pages 32-33 for explanation of data and symbols.

	Emission Standard	Fuel Economy City	Hwy	Fuel Cost/yr	Health Cost/yr	GHG tons/yr	EDX ¢/mi	Green Score	Class Ranking
COMPACT CARS (cont.)									
MERCURY TRACER (cont.)									
2.0L 4, manual	Tier 1	28	37	$540	$160	9	2.02	**29**	▲
2.0L 4, auto	Tier 1	25	34	$600	$170	10	2.13	**27**	○
NISSAN ALTIMA									
2.4L 4, manual	LEV	24	31	$640	$150	10	2.07	**28**	▲
2.4L 4, auto	LEV	22	30	$690	$160	11	2.15	**27**	○
2.4L 4, manual	Tier 1	24	31	$640	$180	10	2.23	**25**	○
2.4L 4, auto	Tier 1	22	30	$690	$180	11	2.30	**24**	○
OLDSMOBILE ALERO									
2.4L 4, auto	LEV	22	30	$690	$160	11	2.21	**26**	○
2.4L 4, auto	TLEV	22	30	$690	$180	11	2.34	**24**	○
3.3L 6, auto	LEV	20	28	$750	$170	12	2.34	**24**	○
2.4L 4, auto	Tier 1	22	30	$690	$180	11	2.36	**24**	○
3.3L 6, auto	TLEV	20	28	$750	$190	12	2.47	**22**	▽
3.3L 6, auto	Tier 1	20	28	$750	$190	12	2.50	**22**	▽
PLYMOUTH NEON									
2.0L 4, manual	LEV	28	39	$540	$140	9	1.84	**32**	✔
2.0L 4, manual	Tier 1	28	39	$540	$160	9	2.00	**29**	▲
2.0L 4, auto	LEV	23	32	$640	$150	10	2.05	**28**	▲
2.0L 4, auto	Tier 1	23	32	$640	$170	10	2.21	**26**	○
PONTIAC GRAND AM									
2.4L 4, auto	LEV	22	30	$690	$160	11	2.21	**26**	○
2.4L 4, auto	TLEV	22	30	$690	$180	11	2.34	**24**	○
3.3L 6, auto	LEV	20	28	$750	$170	12	2.34	**24**	○
2.4L 4, auto	Tier 1	22	30	$690	$180	11	2.36	**24**	○
3.3L 6, auto	TLEV	20	28	$750	$190	12	2.47	**22**	▽
3.3L 6, auto	Tier 1	20	28	$750	$190	12	2.50	**22**	▽
SATURN SL									
1.9L 4, manual	LEV	29	40	$520	$140	8	1.79	**33**	✔
1.9L 4, auto	LEV	27	37	$570	$140	9	1.88	**31**	✔
1.9L 4, manual	LEV	27	38	$560	$140	9	1.89	**31**	✔
1.9L 4, manual	TLEV	29	40	$520	$160	8	1.92	**30**	▲
1.9L 4, manual	Tier 1	29	40	$520	$160	8	1.94	**30**	▲
1.9L 4, auto	LEV	25	35	$600	$150	9	1.96	**30**	▲
1.9L 4, auto	TLEV	27	37	$570	$160	9	2.01	**29**	▲
1.9L 4, manual	TLEV	27	38	$560	$160	9	2.02	**29**	▲
1.9L 4, auto	Tier 1	27	37	$570	$160	9	2.03	**28**	▲

See pages 32-33 for explanation of data and symbols.

	Emission Standard	Fuel Economy City	Fuel Economy Hwy	Fuel Cost/yr	Health Cost/yr	GHG tons/yr	EDX ¢/mi	Green Score	Class Ranking
COMPACT CARS (cont.)									
SATURN SL (cont.)									
1.9L 4, manual	Tier 1	27	38	$560	$170	9	2.04	**28**	▲
1.9L 4, auto	TLEV	25	35	$600	$170	10	2.09	**27**	▲
1.9L 4, auto	Tier 1	25	35	$600	$170	10	2.12	**27**	○
SUBARU LEGACY AWD									
2.2L 4, manual	LEV	22	29	$690	$160	11	2.15	**26**	○
2.2L 4, auto	LEV	22	29	$690	$160	11	2.16	**26**	○
2.2L 4, manual	TLEV	22	29	$690	$180	11	2.29	**25**	○
2.2L 4, auto	TLEV	22	29	$690	$180	11	2.29	**25**	○
2.2L 4, manual	Tier 1	22	29	$690	$180	11	2.31	**24**	○
2.2L 4, auto	Tier 1	22	29	$690	$180	11	2.32	**24**	○
2.5L 4, auto	LEV	21	26	$750	$170	12	2.33	**24**	○
2.5L 4, manual	LEV	21	27	$750	$170	12	2.33	**24**	○
2.5L 4, auto	TLEV	21	26	$750	$190	12	2.46	**22**	▽
2.5L 4, manual	TLEV	21	27	$750	$190	12	2.46	**22**	▽
2.5L 4, auto	Tier 1	21	26	$750	$190	12	2.48	**22**	▽
2.5L 4, manual	Tier 1	21	27	$750	$190	12	2.49	**22**	▽
TOYOTA CAMRY SOLARA									
2.2L 4, manual	LEV	23	32	$640	$160	10	2.14	**27**	○
2.2L 4, auto	LEV	23	30	$660	$160	11	2.17	**26**	○
3.0L 6, manual	LEV	21	28	$720	$160	11	2.29	**25**	○
2.2L 4, manual	Tier 1	23	32	$640	$180	11	2.30	**24**	○
2.2L 4, auto	Tier 1	23	30	$660	$180	11	2.33	**24**	○
3.0L 6, auto	LEV	20	28	$750	$170	12	2.35	**24**	○
3.0L 6, manual	Tier 1	21	28	$720	$190	11	2.44	**23**	▽
3.0L 6, auto	Tier 1	20	28	$750	$190	12	2.51	**22**	✖
TOYOTA COROLLA									
1.8L 4, manual	TLEV	31	38	$510	$160	8	1.92	**30**	▲
1.8L 4, manual	Tier 1	31	38	$510	$160	8	1.95	**30**	▲
1.8L 4, auto 4 sp	TLEV	28	36	$560	$160	9	1.99	**29**	▲
1.8L 4, auto 4 sp	Tier 1	28	36	$560	$160	9	2.02	**29**	▲
1.8L 4, auto 3 sp	TLEV	28	33	$570	$160	9	2.04	**28**	▲
1.8L 4, auto 3 sp	Tier 1	28	33	$570	$170	9	2.07	**28**	▲
VOLKSWAGEN GOLF									
2.0L 4, manual	TLEV	24	31	$640	$170	10	2.21	**26**	○
2.0L 4, manual	Tier 1	24	31	$640	$180	10	2.24	**25**	○

See pages 32-33 for explanation of data and symbols.

	Emission Standard	Fuel Economy City	Hwy	Fuel Cost/yr	Health Cost/yr	GHG tons/yr	EDX ¢/mi	Green Score	Class Ranking
COMPACT CARS (cont.)									
VOLKSWAGEN GOLF (cont.)									
2.0L 4, auto	TLEV	22	28	$720	$180	11	2.34	**24**	○
2.0L 4, auto	Tier 1	22	28	$720	$180	11	2.36	**24**	○
VOLKSWAGEN GTI									
2.0L 4, manual	TLEV	24	32	$640	$170	10	2.15	**26**	○
2.0L 4, manual	Tier 1	24	32	$640	$170	10	2.18	**26**	○
2.0L 4, auto	TLEV	22	28	$720	$180	11	2.34	**24**	○
2.0L 4, auto	Tier 1	22	28	$720	$180	11	2.36	**24**	○
2.8L 6, manual	TLEV	19	26	$790	$190	12	2.50	**22**	▽
2.8L 6, manual	Tier 1	19	26	$790	$190	12	2.52	**22**	✖
VOLKSWAGEN JETTA									
2.0L 4, manual	TLEV	24	31	$640	$170	10	2.21	**26**	○
2.0L 4, manual	Tier 1	24	31	$640	$180	10	2.24	**25**	○
2.0L 4, auto	TLEV	22	28	$720	$180	11	2.34	**24**	○
2.0L 4, auto	Tier 1	22	28	$720	$180	11	2.36	**24**	○
1.9L 4, manual [D]	Tier 1-D	40	49	$380	$260	6	2.39	**23**	○
2.8L 6, manual	TLEV	19	26	$790	$190	12	2.50	**22**	▽
2.8L 6, manual	Tier 1	19	26	$790	$190	12	2.52	**22**	✖
2.8L 6, auto	TLEV	18	25	$820	$190	13	2.55	**21**	✖
2.8L 6, auto	Tier 1	18	25	$820	$190	13	2.58	**21**	✖
VOLKSWAGEN NEW GOLF									
2.0L 4, manual	TLEV	24	31	$640	$170	10	2.21	**26**	○
2.0L 4, manual	Tier 1	24	31	$640	$180	10	2.23	**25**	○
2.0L 4, auto	TLEV	22	28	$720	$180	11	2.32	**24**	○
2.0L 4, auto	Tier 1	22	28	$720	$180	11	2.34	**24**	○
1.9L 4, manual [D]	Tier 1-D	42	49	$370	$260	6	2.37	**24**	○
1.9L 4, auto [D]	Tier 1-D	34	45	$430	$270	7	2.49	**22**	▽
VOLKSWAGEN NEW GTI									
2.0L 4, manual	TLEV	24	31	$640	$170	10	2.21	**26**	○
2.0L 4, manual	Tier 1	24	31	$640	$180	10	2.23	**25**	○
2.0L 4, auto	TLEV	22	28	$720	$180	11	2.32	**24**	○
2.0L 4, auto	Tier 1	22	28	$720	$180	11	2.34	**24**	○
2.8L 6, manual	TLEV	20	28	$750	$180	11	2.40	**23**	○
2.8L 6, manual	Tier 1	20	28	$750	$180	11	2.42	**23**	▽
VOLKSWAGEN NEW JETTA									
2.0L 4, manual	Tier 1	24	31	$640	$180	10	2.23	**25**	○
2.0L 4, auto	Tier 1	22	28	$720	$180	11	2.34	**24**	○

See pages 32-33 for explanation of data and symbols.

	Emission Standard	Fuel Economy City	Fuel Economy Hwy	Fuel Cost/yr	Health Cost/yr	GHG tons/yr	EDX ¢/mi	Green Score	Class Ranking
COMPACT CARS (cont.)									
VOLKSWAGEN NEW JETTA (cont.)									
1.9L 4, manual [D]	Tier 1-D	42	49	$370	$260	6	2.37	**24**	○
1.9L 4, auto [D]	Tier 1-D	34	45	$430	$270	7	2.49	**22**	▽
2.8L 6, manual	Tier 1	19	28	$750	$190	12	2.52	**22**	✖
2.8L 6, auto	Tier 1	19	26	$820	$200	13	2.60	**21**	✖
VOLVO C70									
2.3L 5, manual [P]	Tier 1	20	28	$850	$190	12	2.50	**22**	▽
2.3L 5, auto [P]	Tier 1	20	27	$850	$190	12	2.51	**22**	✖
2.4L 5, auto [P]	Tier 1	20	27	$890	$190	12	2.52	**22**	✖
SMALL WAGONS									
AUDI A4 AVANT QUATTRO									
1.8L 4, manual [P]	TLEV	21	29	$810	$180	11	2.40	**23**	▽
1.8L 4, manual [P]	Tier 1	21	29	$810	$190	11	2.42	**23**	▽
2.8L 6, manual [P]	TLEV	19	27	$890	$190	12	2.52	**22**	✖
2.8L 6, manual [P]	Tier 1	19	27	$890	$190	12	2.55	**21**	✖
1.8L 4, auto [P]	TLEV	18	27	$930	$190	12	2.56	**21**	✖
1.8L 4, auto [P]	Tier 1	18	27	$930	$190	12	2.59	**21**	✖
2.8L 6, auto [P]	TLEV	17	27	$930	$200	13	2.67	**20**	✖
2.8L 6, auto [P]	Tier 1	17	27	$930	$200	13	2.70	**20**	✖
BMW 528I TOURING									
2.8L 6, manual	Tier 1	18	26	$820	$200	13	2.67	**20**	✖
2.8L 6, auto	Tier 1	18	26	$820	$200	13	2.68	**20**	✖
BMW 540I TOURING									
4.4L 8, auto	Tier 1	15	21	$960	$220	15	2.98	**17**	✖
DAEWOO NUBIRA STATION WAGON									
2.0L 4, manual	Tier 1	22	31	$660	$180	11	2.27	**25**	▽
2.0L 4, auto	Tier 1	22	30	$690	$180	11	2.30	**24**	▽
FORD ESCORT WAGON									
2.0L 4, manual	LEV	28	37	$540	$140	9	1.86	**31**	✔
2.0L 4, auto	LEV	25	34	$600	$150	10	1.97	**29**	▲
2.0L 4, manual	Tier 1	28	37	$540	$160	9	2.02	**29**	○
2.0L 4, auto	Tier 1	25	34	$600	$170	10	2.13	**27**	○

See pages 32-33 for explanation of data and symbols.

	Emission Standard	Fuel Economy City	Hwy	Fuel Cost/yr	Health Cost/yr	GHG tons/yr	EDX ¢/mi	Green Score	Class Ranking
SMALL WAGONS (cont.)									
HYUNDAI ELANTRA WAGON									
2.0L 4, manual	LEV	23	32	$640	$150	10	2.07	**28**	○
2.0L 4, auto	LEV	21	30	$720	$160	11	2.20	**26**	○
2.0L 4, manual	Tier 1	23	32	$640	$180	10	2.23	**25**	▽
2.0L 4, auto	Tier 1	21	30	$720	$180	11	2.35	**24**	▽
MERCURY TRACER WAGON									
2.0L 4, manual	LEV	28	37	$540	$140	9	1.86	**31**	✔
2.0L 4, auto	LEV	25	34	$600	$150	10	1.97	**29**	▲
2.0L 4, manual	Tier 1	28	37	$540	$160	9	2.02	**29**	○
2.0L 4, auto	Tier 1	25	34	$600	$170	10	2.13	**27**	○
SATURN SW									
1.9L 4, manual	LEV	28	38	$560	$140	9	1.84	**32**	✔
1.9L 4, manual	LEV	27	38	$560	$140	9	1.89	**31**	✔
1.9L 4, auto	LEV	26	35	$570	$140	9	1.93	**30**	▲
1.9L 4, auto	LEV	25	35	$600	$150	9	1.96	**30**	▲
1.9L 4, manual	TLEV	28	38	$560	$160	9	1.97	**29**	▲
1.9L 4, manual	Tier 1	28	38	$560	$160	9	1.99	**29**	▲
1.9L 4, manual	TLEV	27	38	$560	$160	9	2.02	**29**	○
1.9L 4, manual	Tier 1	27	38	$560	$170	9	2.04	**28**	○
1.9L 4, auto	TLEV	26	35	$570	$160	9	2.06	**28**	○
1.9L 4, auto	Tier 1	26	35	$570	$170	9	2.09	**27**	○
1.9L 4, auto	TLEV	25	35	$600	$170	10	2.09	**27**	○
1.9L 4, auto	Tier 1	25	35	$600	$170	10	2.12	**27**	○
SUBARU IMPREZA WAGON									
2.2L 4, auto	LEV	23	29	$690	$160	11	2.15	**27**	○
2.2L 4, manual	LEV	22	29	$690	$160	11	2.15	**26**	○
2.2L 4, auto	TLEV	23	29	$690	$180	11	2.28	**25**	▽
2.2L 4, manual	TLEV	22	29	$690	$180	11	2.29	**25**	▽
2.2L 4, auto	Tier 1	23	29	$690	$180	11	2.31	**24**	▽
2.2L 4, manual	Tier 1	22	29	$690	$180	11	2.31	**24**	▽
SUZUKI ESTEEM WAGON									
1.6L 4, manual	TLEV	30	36	$540	$160	9	1.96	**29**	▲
1.6L 4, manual	Tier 1	30	36	$540	$160	9	1.99	**29**	▲
1.6L 4, auto	TLEV	26	33	$600	$170	10	2.09	**27**	○
1.6L 4, auto	Tier 1	26	33	$600	$170	10	2.12	**27**	○

See pages 32-33 for explanation of data and symbols.

	Emission Standard	Fuel Economy City	Fuel Economy Hwy	Fuel Cost/yr	Health Cost/yr	GHG tons/yr	EDX ¢/mi	Green Score	Class Ranking
MIDSIZE CARS									
ACURA 3.2TL									
3.2L 6, auto [P]	Tier 1	19	27	$890	$190	12	2.57	**21**	▽
ACURA 3.5RL									
3.5L 6, auto [P]	Tier 1	18	24	$980	$200	13	2.74	**19**	✖
AUDI A6									
2.8L 6, auto [P]	TLEV	17	27	$930	$200	13	2.66	**20**	▽
2.8L 6, auto [P]	Tier 1	17	27	$930	$200	13	2.69	**20**	▽
AUDI A6 QUATTRO									
2.8L 6, auto [P]	TLEV	17	26	$980	$200	13	2.73	**19**	✖
2.8L 6, auto [P]	Tier 1	17	26	$980	$210	13	2.76	**19**	✖
AUDI A8									
3.7L 8, auto [P]	Tier 1	17	26	$980	$200	13	2.75	**19**	✖
AUDI A8 QUATTRO									
4.2L 8, auto [P]	Tier 1	17	25	$980	$210	14	2.79	**19**	✖
BENTLEY ARNAGE									
4.4L 8, auto [P]	Tier 1	12	16	$1,390	$250	19	3.61	**12**	✖
BENTLEY TURBO RT									
6.8L 8, auto [P]	Tier 1	11	16	$1,500	$260	20	3.83	**11**	✖
BMW 740I									
4.4L 8, auto	Tier 1	17	23	$910	$210	14	2.86	**18**	✖
4.4L 8, auto	Tier 1	15	21	$1,010	$220	15	3.06	**16**	✖
BUICK CENTURY									
3.1L 6, auto	TLEV	20	29	$750	$190	12	2.46	**22**	○
3.1L 6, auto	Tier 1	20	29	$750	$190	12	2.48	**22**	○
BUICK REGAL									
3.8L 6, auto	TLEV	19	30	$750	$190	12	2.46	**22**	○
3.8L 6, auto	Tier 1	19	30	$750	$190	12	2.49	**22**	○
3.8L 6, auto [P]	TLEV	18	27	$930	$200	13	2.66	**20**	▽
3.8L 6, auto [P]	Tier 1	18	27	$930	$200	13	2.69	**20**	▽
BUICK RIVIERA									
3.8L 6, auto [P]	Tier 1	18	27	$930	$200	13	2.66	**20**	▽
CADILLAC CATERA									
3.0L 6, auto [P]	Tier 1	18	24	$980	$200	13	2.72	**19**	✖

See pages 32-33 for explanation of data and symbols.

	Emission Standard	Fuel Economy City	Fuel Economy Hwy	Fuel Cost/yr	Health Cost/yr	GHG tons/yr	EDX ¢/mi	Green Score	Class Ranking
MIDSIZE CARS (cont.)									
CADILLAC ELDORADO									
4.6L 8, auto [P]	Tier 1	17	26	$980	$200	13	2.75	**19**	✖
CADILLAC SEVILLE									
4.6L 8, auto [P]	Tier 1	17	26	$980	$200	13	2.75	**19**	✖
CHEVROLET LUMINA/MONTE CARLO									
3.1L 6, auto	TLEV	20	29	$750	$190	12	2.46	**22**	○
3.8L 6, auto	TLEV	19	30	$750	$190	12	2.46	**22**	○
3.1L 6, auto	Tier 1	20	29	$750	$190	12	2.48	**22**	○
3.8L 6, auto	Tier 1	19	30	$750	$190	12	2.49	**22**	○
CHEVROLET MALIBU									
2.4L 4, auto	LEV	22	30	$690	$160	11	2.21	**26**	▲
3.1L 6, auto	LEV	20	29	$750	$170	12	2.33	**24**	○
2.4L 4, auto	Tier 1	22	30	$690	$180	11	2.36	**24**	○
3.1L 6, auto	Tier 1	20	29	$750	$190	12	2.48	**22**	○
CHRYSLER CIRRUS									
2.5L 6, auto	LEV	19	27	$790	$170	12	2.40	**23**	○
2.5L 6, auto	TLEV	19	27	$790	$190	12	2.53	**22**	▽
2.5L 6, auto	Tier 1	19	27	$790	$190	12	2.56	**21**	▽
DAEWOO LEGANZA									
2.2L 4, manual	TLEV	20	29	$720	$180	12	2.43	**23**	○
2.2L 4, manual	Tier 1	20	29	$720	$190	12	2.45	**22**	○
2.2L 4, auto	TLEV	20	28	$750	$190	12	2.47	**22**	○
2.2L 4, auto	Tier 1	20	28	$750	$190	12	2.50	**22**	○
DODGE STRATUS									
2.0L 4, manual	LEV	26	37	$570	$150	9	1.93	**30**	✔
2.0L 4, manual	TLEV	26	37	$570	$170	9	2.06	**28**	✔
2.0L 4, manual	Tier 1	26	37	$570	$170	9	2.09	**27**	✔
2.4L 4, auto	LEV	21	30	$720	$160	11	2.25	**25**	▲
2.4L 4, auto	TLEV	21	30	$720	$180	11	2.38	**23**	○
2.5L 6, auto	LEV	19	27	$790	$170	12	2.40	**23**	○
2.4L 4, auto	Tier 1	21	30	$720	$190	11	2.41	**23**	○
2.5L 6, auto	TLEV	19	27	$790	$190	12	2.53	**22**	▽
2.5L 6, auto	Tier 1	19	27	$790	$190	12	2.56	**21**	▽
FORD TAURUS[6]									
2.9L 6, auto	LEV	20	28	$750	$170	12	2.35	**24**	○
3.0L 6, auto	LEV	19	27	$790	$170	12	2.39	**23**	○

[6] *An ethanol-gasoline flex-fuel version of this vehicle is also available.*

	Emission Standard	Fuel Economy City	Hwy	Fuel Cost/yr	Health Cost/yr	GHG tons/yr	EDX ¢/mi	Green Score	Class Ranking
MIDSIZE CARS (cont.)									
FORD TAURUS[6] (cont.)									
3.0L 6, auto	LEV	18	26	$820	$170	13	2.48	**22**	○
2.9L 6, auto	Tier 1	20	28	$750	$190	12	2.51	**22**	▽
3.0L 6, auto	Tier 1	19	27	$790	$190	12	2.54	**21**	▽
3.0L 6, auto	Tier 1	18	26	$820	$200	13	2.64	**20**	▽
3.4L 8, auto [P]	LEV	16	25	$1,030	$180	14	2.64	**20**	▽
3.4L 8, auto [P]	Tier 1	16	25	$1,030	$210	14	2.80	**19**	✖
HONDA ACCORD									
2.3L 4, manual	LEV	25	31	$640	$150	10	2.06	**28**	✔
2.3L 4, auto	ULEV	23	30	$660	$150	11	2.12	**27**	▲
2.3L 4, auto	LEV	22	29	$690	$160	11	2.16	**26**	▲
2.3L 4, auto	LEV	23	30	$660	$160	11	2.18	**26**	▲
2.3L 4, manual	Tier 1	25	31	$640	$170	10	2.21	**26**	▲
3.0L 6, auto	LEV	20	28	$750	$170	12	2.31	**24**	○
2.3L 4, auto	Tier 1	22	29	$690	$180	11	2.32	**24**	○
2.3L 4, auto	Tier 1	23	30	$660	$180	11	2.34	**24**	○
3.0L 6, auto	Tier 1	20	28	$750	$190	12	2.47	**22**	○
HYUNDAI SONATA									
2.4L 4, manual	TLEV	21	30	$690	$180	11	2.37	**23**	○
2.4L 4, manual	Tier 1	21	30	$690	$190	11	2.40	**23**	○
2.4L 4, auto	TLEV	21	28	$720	$180	12	2.43	**23**	○
2.5L 6, manual	TLEV	20	29	$750	$180	12	2.43	**23**	○
2.4L 4, auto	Tier 1	21	28	$720	$190	12	2.45	**22**	○
2.5L 6, manual	Tier 1	20	29	$750	$190	12	2.46	**22**	○
2.5L 6, auto	TLEV	20	28	$750	$190	12	2.46	**22**	○
2.5L 6, auto	Tier 1	20	28	$750	$190	12	2.49	**22**	○
INFINITI I30									
3.0L 6, manual	TLEV	21	26	$750	$180	12	2.43	**23**	○
3.0L 6, auto	TLEV	21	28	$750	$190	12	2.44	**23**	○
3.0L 6, manual	Tier 1	21	26	$750	$190	12	2.45	**22**	○
3.0L 6, auto	Tier 1	21	28	$750	$190	12	2.47	**22**	○
INFINITI Q45									
4.1L 8, auto [P]	Tier 1	17	24	$980	$210	13	2.77	**19**	✖
JAGUAR VANDEN PLAS									
4.0L 8, auto [P]	Tier 1	16	22	$1,080	$220	15	2.97	**17**	✖

[6] *An ethanol-gasoline flex-fuel version of this vehicle is also available.*

	Emission Standard	Fuel Economy City	Hwy	Fuel Cost/yr	Health Cost/yr	GHG tons/yr	EDX ¢/mi	Green Score	Class Ranking
MIDSIZE CARS (cont.)									
LEXUS GS 300/GS 400									
3.0L 6, auto [P]	TLEV	20	25	$890	$190	12	2.58	**21**	▽
3.0L 6, auto [P]	Tier 1	20	25	$890	$200	12	2.61	**21**	▽
4.0L 8, auto [P]	TLEV	17	24	$980	$200	13	2.73	**19**	✖
4.0L 8, auto [P]	Tier 1	17	24	$980	$210	13	2.76	**19**	✖
LEXUS LS 400									
4.0L 8, auto [P]	TLEV	18	25	$930	$200	13	2.66	**20**	▽
4.0L 8, auto [P]	Tier 1	18	25	$930	$200	13	2.69	**20**	▽
MAZDA 626									
2.0L 4, manual	TLEV	26	33	$600	$170	10	2.10	**27**	▲
2.0L 4, manual	Tier 1	26	33	$600	$170	10	2.13	**27**	▲
2.0L 4, auto	TLEV	22	29	$720	$180	11	2.32	**24**	○
2.0L 4, auto	Tier 1	22	29	$720	$180	11	2.34	**24**	○
2.5L 6, manual [P]	TLEV	21	27	$810	$180	11	2.36	**24**	○
2.5L 6, manual [P]	Tier 1	21	27	$810	$180	11	2.39	**23**	○
2.5L 6, auto [P]	TLEV	20	26	$850	$190	12	2.48	**22**	○
2.5L 6, auto [P]	Tier 1	20	26	$850	$190	12	2.51	**22**	▽
MERCEDES E300 TURBO DIESEL									
3.0L 6, auto [D]	Tier 1-D	26	36	$550	$290	9	2.83	**18**	✖
MERCEDES E320									
3.2L 6, auto [P]	LEV	21	30	$810	$170	12	2.33	**24**	○
3.2L 6, auto [P]	Tier 1	21	30	$810	$190	12	2.49	**22**	○
MERCEDES E320 4MATIC									
3.2L 6, auto [P]	LEV	20	28	$850	$170	12	2.38	**23**	○
3.2L 6, auto [P]	Tier 1	20	28	$850	$200	12	2.54	**21**	▽
MERCEDES E430									
4.3L 8, auto [P]	LEV	19	26	$930	$180	13	2.50	**22**	▽
4.3L 8, auto [P]	Tier 1	19	26	$930	$200	13	2.66	**20**	▽
MERCURY SABLE									
2.9L 6, auto	LEV	20	28	$750	$170	12	2.35	**24**	○
3.0L 6, auto	LEV	18	26	$820	$170	13	2.48	**22**	○
2.9L 6, auto	Tier 1	20	28	$750	$190	12	2.51	**22**	▽
3.0L 6, auto	Tier 1	18	26	$820	$200	13	2.64	**20**	▽
MITSUBISHI DIAMANTE SEDAN									
3.5L 6, auto [P]	LEV	18	24	$980	$180	13	2.57	**21**	▽
3.5L 6, auto [P]	Tier 1	18	24	$980	$200	13	2.73	**19**	✖

See pages 32-33 for explanation of data and symbols.

	Emission Standard	Fuel Economy City	Fuel Economy Hwy	Fuel Cost/yr	Health Cost/yr	GHG tons/yr	EDX ¢/mi	Green Score	Class Ranking
MIDSIZE CARS (cont.)									
MITSUBISHI GALANT									
2.3L 4, manual	LEV	23	31	$660	$150	10	2.11	**27**	▲
2.3L 4, auto	LEV	21	28	$720	$160	11	2.23	**25**	▲
2.3L 4, manual	Tier 1	23	31	$660	$180	10	2.27	**25**	▲
3.0L 6, auto	LEV	20	27	$790	$170	12	2.36	**24**	○
2.3L 4, auto	Tier 1	21	28	$720	$180	11	2.39	**23**	○
3.0L 6, auto	Tier 1	20	27	$790	$190	12	2.52	**22**	▽
NISSAN MAXIMA									
3.0L 6, manual	TLEV	22	27	$720	$180	11	2.42	**23**	○
3.0L 6, auto	TLEV	21	28	$750	$180	12	2.43	**23**	○
3.0L 6, manual	Tier 1	22	27	$720	$190	12	2.45	**23**	○
3.0L 6, auto	Tier 1	21	28	$750	$190	12	2.46	**22**	○
OLDSMOBILE AURORA									
4.0L 8, auto [P]	Tier 1	17	26	$980	$200	13	2.74	**19**	✖
OLDSMOBILE CUTLASS									
2.4L 4, auto	LEV	22	30	$690	$160	11	2.21	**26**	▲
3.1L 6, auto	LEV	20	29	$750	$170	12	2.33	**24**	○
2.4L 4, auto	TLEV	22	30	$690	$180	11	2.34	**24**	○
2.4L 4, auto	Tier 1	22	30	$690	$180	11	2.36	**24**	○
3.1L 6, auto	TLEV	20	29	$750	$190	12	2.46	**22**	○
3.1L 6, auto	Tier 1	20	29	$750	$190	12	2.48	**22**	○
OLDSMOBILE INTRIGUE									
3.8L 6, auto	TLEV	19	30	$750	$190	12	2.46	**22**	○
3.8L 6, auto	Tier 1	19	30	$750	$190	12	2.49	**22**	○
3.5L 6, auto	TLEV	19	27	$790	$190	12	2.55	**21**	▽
3.5L 6, auto	Tier 1	19	27	$790	$190	12	2.58	**21**	▽
PLYMOUTH BREEZE									
2.0L 4, manual	LEV	26	37	$570	$150	9	1.93	**30**	✔
2.0L 4, manual	TLEV	26	37	$570	$170	9	2.06	**28**	✔
2.0L 4, manual	Tier 1	26	37	$570	$170	9	2.09	**27**	✔
2.0L 4, auto	LEV	22	31	$690	$160	11	2.21	**26**	▲
2.4L 4, auto	LEV	21	30	$720	$160	11	2.25	**25**	▲
2.0L 4, auto	TLEV	22	31	$690	$180	11	2.34	**24**	○
2.0L 4, auto	Tier 1	22	31	$690	$180	11	2.36	**24**	○
2.4L 4, auto	TLEV	21	30	$720	$180	11	2.38	**23**	○
2.4L 4, auto	Tier 1	21	30	$720	$190	11	2.41	**23**	○

See pages 32-33 for explanation of data and symbols.

	Emission Standard	Fuel Economy City	Fuel Economy Hwy	Fuel Cost/yr	Health Cost/yr	GHG tons/yr	EDX ¢/mi	Green Score	Class Ranking
MIDSIZE CARS (cont.)									
PONTIAC GRAND PRIX									
3.1L 6, auto	TLEV	20	29	$750	$190	12	2.46	**22**	○
3.8L 6, auto	TLEV	19	30	$750	$190	12	2.46	**22**	○
3.1L 6, auto	Tier 1	20	29	$750	$190	12	2.48	**22**	○
3.8L 6, auto	Tier 1	19	30	$750	$190	12	2.49	**22**	○
3.8L 6, auto [P]	TLEV	18	28	$930	$190	13	2.59	**21**	▽
3.8L 6, auto [P]	Tier 1	18	28	$930	$200	13	2.62	**21**	▽
ROLLS-ROYCE SILVER SERAPH									
5.4L 2, auto 5 sp [P]	Tier 1	12	16	$1,390	$250	19	3.58	**13**	✖
ROLLS-ROYCE SILVER SPUR									
6.8L 8, auto 4 sp [P]	Tier 1	11	16	$1,500	$260	20	3.83	**11**	✖
SAAB 9-3									
2.0L 4, manual	Tier 1	20	27	$750	$190	12	2.49	**22**	○
2.0L 4, manual [P]	Tier 1	19	27	$890	$190	12	2.54	**21**	▽
2.0L 4, auto	Tier 1	19	25	$820	$200	13	2.62	**21**	▽
SAAB 9-5									
2.3L 4, manual	Tier 1	21	30	$720	$190	11	2.46	**22**	○
2.3L 4, auto	Tier 1	19	26	$820	$200	13	2.66	**20**	▽
3.0L 6, auto	Tier 1	18	26	$820	$200	13	2.68	**20**	▽
TOYOTA CAMRY									
2.2L 4, manual	LEV	23	32	$640	$160	10	2.14	**27**	▲
2.2L 4, auto	LEV	23	30	$660	$160	11	2.17	**26**	▲
3.0L 6, manual	LEV	21	28	$720	$160	11	2.29	**25**	▲
2.2L 4, manual	Tier 1	23	32	$640	$180	11	2.30	**24**	▲
2.2L 4, auto	Tier 1	23	30	$660	$180	11	2.33	**24**	○
3.0L 6, auto	LEV	20	28	$750	$170	12	2.35	**24**	○
3.0L 6, manual	Tier 1	21	28	$720	$190	11	2.44	**23**	○
3.0L 6, auto	Tier 1	20	28	$750	$190	12	2.51	**22**	▽
VOLKSWAGEN PASSAT									
1.8L 4, manual [P]	TLEV	23	32	$750	$180	10	2.27	**25**	▲
1.8L 4, manual [P]	Tier 1	23	32	$750	$180	11	2.30	**24**	▲
1.8L 4, auto [P]	TLEV	21	31	$810	$180	11	2.39	**23**	○
1.8L 4, auto [P]	Tier 1	21	31	$810	$190	11	2.41	**23**	○
2.8L 6, manual [P]	TLEV	20	29	$850	$190	12	2.45	**22**	○
2.8L 6, manual [P]	Tier 1	20	29	$850	$190	12	2.48	**22**	○
2.8L 6, auto [P]	TLEV	18	29	$930	$190	13	2.58	**21**	▽
2.8L 6, auto [P]	Tier 1	18	29	$930	$200	13	2.60	**21**	▽

See pages 32-33 for explanation of data and symbols.

	Emission Standard	Fuel Economy City	Fuel Economy Hwy	Fuel Cost/yr	Health Cost/yr	GHG tons/yr	EDX ¢/mi	Green Score	Class Ranking
MIDSIZE CARS (cont.)									
VOLKSWAGEN PASSAT SYNCRO									
2.8L 6, auto [P]	Tier 1	17	26	$980	$200	13	2.73	**19**	✖
VOLVO S70[7]									
2.4L 5, manual [P]	Tier 1	21	29	$810	$190	11	2.43	**23**	○
2.4L 5, auto [P]	Tier 1	20	28	$850	$190	12	2.47	**22**	○
2.3L 5, manual [P]	Tier 1	20	28	$850	$190	12	2.50	**22**	○
2.3L 5, auto [P]	Tier 1	20	27	$850	$190	12	2.51	**22**	▽
2.4L 5, auto [P]	Tier 1	20	27	$890	$190	12	2.52	**22**	▽
2.4L 5, auto [P] Awd	Tier 1	18	25	$930	$200	13	2.69	**20**	▽
2.3L 5, auto [P] Awd	Tier 1	18	25	$930	$200	13	2.70	**20**	✖
VOLVO S80[8]									
2.9L 6, auto [P]	Tier 1	19	27	$890	$200	12	2.60	**21**	▽
2.8L 6, auto [P]	Tier 1	18	27	$890	$200	13	2.63	**20**	▽

	Emission Standard	Fuel Economy City	Fuel Economy Hwy	Fuel Cost/yr	Health Cost/yr	GHG tons/yr	EDX ¢/mi	Green Score	Class Ranking
MIDSIZE WAGONS									
AUDI A6 AVANT QUATTRO									
2.8L 6, auto [P]	TLEV	17	26	$980	$200	13	2.73	**19**	✖
2.8L 6, auto [P]	Tier 1	17	26	$980	$210	13	2.76	**19**	✖
FORD TAURUS WAGON									
3.0L 6, auto	LEV	18	27	$820	$170	12	2.43	**23**	○
3.0L 6, auto	LEV	18	26	$820	$180	13	2.54	**21**	○
3.0L 6, auto	Tier 1	18	27	$820	$190	12	2.59	**21**	▽
3.0L 6, auto	Tier 1	18	26	$820	$200	13	2.70	**20**	✖
ISUZU OASIS									
2.3L 4, auto	Tier 1	21	26	$750	$190	12	2.48	**22**	○
LEXUS RX 300									
3.0L 6, auto	LEV	19	24	$820	$180	13	2.53	**22**	○
3.0L 6, auto 4 wd	LEV	19	22	$860	$180	13	2.59	**21**	▽
3.0L 6, auto	Tier 1	19	24	$820	$200	13	2.68	**20**	✖
3.0L 6, auto 4 wd	Tier 1	19	22	$860	$200	13	2.75	**19**	✖
MERCEDES E320									
3.2L 6, auto [P]	LEV	20	28	$850	$170	12	2.39	**23**	○
3.2L 6, auto [P]	Tier 1	20	28	$850	$200	12	2.54	**21**	○

[7/8] *A CNG-gasoline bi-fuel version of this vehicle is also available.*

	Emission Standard	Fuel Economy City	Fuel Economy Hwy	Fuel Cost/yr	Health Cost/yr	GHG tons/yr	EDX ¢/mi	Green Score	Class Ranking
MIDSIZE WAGONS (cont.)									
MERCEDES E320 4MATIC									
3.2L 6, auto [P]	LEV	20	26	$890	$170	12	2.44	**23**	○
3.2L 6, auto [P]	Tier 1	20	26	$890	$200	12	2.60	**21**	▽
MERCURY SABLE WAGON									
3.0L 6, auto	LEV	19	27	$790	$170	12	2.40	**23**	○
3.0L 6, auto	LEV	18	26	$820	$180	13	2.54	**21**	○
3.0L 6, auto	Tier 1	19	27	$790	$190	12	2.56	**21**	▽
3.0L 6, auto	Tier 1	18	26	$820	$200	13	2.70	**20**	✖
NISSAN ALTRA									
Electric[9]	ZEV	2.9	3.3	$480	$70	5	1.00	**52**	✔
SUBARU LEGACY WAGON									
2.2L 4, manual	LEV	22	29	$690	$160	11	2.15	**26**	✔
2.2L 4, auto	LEV	22	29	$690	$160	11	2.22	**26**	▲
2.2L 4, manual	TLEV	22	29	$690	$180	11	2.29	**25**	▲
2.2L 4, manual	Tier 1	22	29	$690	$180	11	2.31	**24**	▲
2.5L 4, auto	LEV	21	26	$750	$170	12	2.33	**24**	▲
2.5L 4, manual	LEV	21	27	$750	$170	12	2.33	**24**	▲
2.2L 4, auto	TLEV	22	29	$690	$180	11	2.35	**24**	▲
2.2L 4, auto	Tier 1	22	29	$690	$190	11	2.38	**23**	○
2.5L 4, auto	TLEV	21	26	$750	$190	12	2.46	**22**	○
2.5L 4, manual	TLEV	21	27	$750	$190	12	2.46	**22**	○
2.5L 4, auto	Tier 1	21	26	$750	$190	12	2.48	**22**	○
2.5L 4, manual	Tier 1	21	27	$750	$190	12	2.49	**22**	○
VOLKSWAGEN PASSAT WAGON									
1.8L 4, manual [P]	TLEV	23	32	$750	$180	10	2.27	**25**	▲
1.8L 4, manual [P]	Tier 1	23	32	$750	$180	11	2.30	**24**	▲
1.8L 4, auto [P]	TLEV	21	31	$810	$180	11	2.39	**23**	○
1.8L 4, auto [P]	Tier 1	21	31	$810	$190	11	2.41	**23**	○
2.8L 6, manual [P]	TLEV	20	29	$850	$190	12	2.45	**22**	○
2.8L 6, manual [P]	Tier 1	20	29	$850	$190	12	2.48	**22**	○
2.8L 6, auto [P]	TLEV	18	29	$930	$190	13	2.58	**21**	▽
2.8L 6, auto [P]	Tier 1	18	29	$930	$200	13	2.60	**21**	▽
VOLKSWAGEN PASSAT WAGON SYNCRO									
2.8L 6, auto [P]	TLEV	17	26	$980	$200	13	2.70	**20**	✖
2.8L 6, auto [P]	Tier 1	17	26	$980	$200	13	2.73	**19**	✖

[9] *Electric vehicle fuel economy is given in mi/kWh.*

MIDSIZE WAGONS (cont.)

	Emission Standard	Fuel Economy City	Fuel Economy Hwy	Fuel Cost/yr	Health Cost/yr	GHG tons/yr	EDX ¢/mi	Green Score	Class Ranking
VOLVO V70									
2.4L 5, manual [P]	Tier 1	21	29	$810	$190	11	2.43	**23**	○
2.4L 5, auto [P]	Tier 1	20	28	$850	$190	12	2.47	**22**	○
2.3L 5, manual [P]	Tier 1	20	28	$850	$190	12	2.50	**22**	○
2.3L 5, auto [P]	Tier 1	20	27	$850	$190	12	2.51	**22**	○
2.4L 5, auto [P]	Tier 1	20	27	$890	$190	12	2.52	**22**	○
2.4L 5, auto [P] Awd	Tier 1	18	25	$930	$200	13	2.69	**20**	✖
2.3L 5, auto [P] Awd	Tier 1	18	25	$930	$200	13	2.70	**20**	✖

LARGE CARS

	Emission Standard	Fuel Economy City	Fuel Economy Hwy	Fuel Cost/yr	Health Cost/yr	GHG tons/yr	EDX ¢/mi	Green Score	Class Ranking
BENTLEY BROOKLANDS									
6.8L 8, auto [P]	Tier 1	10	15	$1,620	$280	22	4.16	**9**	✖
BMW 740IL									
4.4L 8, auto	Tier 1	17	23	$910	$210	14	2.86	**18**	▽
BMW 750IL									
5.4L 2, auto	Tier 1	13	20	$1,080	$240	17	3.34	**14**	✖
BUICK LESABRE									
3.8L 6, auto	TLEV	19	30	$750	$190	12	2.46	**22**	▲
3.8L 6, auto	Tier 1	19	30	$750	$190	12	2.49	**22**	▲
BUICK PARK AVENUE									
3.8L 6, auto	TLEV	19	28	$790	$190	12	2.58	**21**	○
3.8L 6, auto	Tier 1	19	28	$790	$200	12	2.61	**21**	○
3.8L 6, auto [P]	TLEV	18	27	$930	$200	13	2.63	**20**	○
3.8L 6, auto [P]	Tier 1	18	27	$930	$200	13	2.66	**20**	○
CADILLAC DEVILLE									
4.6L 8, auto [P]	Tier 1	17	26	$980	$200	13	2.75	**19**	▽
CHRYSLER 300 M									
3.5L 6, auto	TLEV	18	27	$820	$200	13	2.64	**20**	○
3.5L 6, auto	Tier 1	18	27	$820	$200	13	2.67	**20**	○
CHRYSLER CONCORDE									
2.7L 6, auto	TLEV	21	30	$720	$180	11	2.39	**23**	▲
2.7L 6, auto	Tier 1	21	30	$720	$190	11	2.42	**23**	▲
3.2L 6, auto	TLEV	19	29	$790	$190	12	2.57	**21**	○
3.2L 6, auto	Tier 1	19	29	$790	$200	12	2.59	**21**	○

See pages 32-33 for explanation of data and symbols.

	Emission Standard	Fuel Economy City	Fuel Economy Hwy	Fuel Cost/yr	Health Cost/yr	GHG tons/yr	EDX ¢/mi	Green Score	Class Ranking
LARGE CARS (cont.)									
CHRYSLER LHS									
3.5L 6, auto	TLEV	18	27	$820	$200	13	2.64	**20**	○
3.5L 6, auto	Tier 1	18	27	$820	$200	13	2.67	**20**	○
DODGE INTREPID									
2.7L 6, auto	TLEV	21	30	$720	$180	11	2.39	**23**	▲
2.7L 6, auto	Tier 1	21	30	$720	$190	11	2.42	**23**	▲
3.2L 6, auto	TLEV	18	28	$820	$200	13	2.62	**21**	○
3.2L 6, auto	Tier 1	18	28	$820	$200	13	2.65	**20**	○
FORD CROWN VICTORIA									
4.6L 8, auto [CNG][10]	ULEV	14	20	$660	$90	13	2.00	**29**	✔
4.6L 8, auto	LEV	17	24	$910	$180	14	2.65	**20**	○
4.6L 8, auto	Tier 1	17	24	$910	$210	14	2.81	**19**	▽
LINCOLN CONTINENTAL									
4.6L 8, auto [P]	TLEV	17	25	$980	$200	13	2.73	**19**	▽
4.6L 8, auto [P]	Tier 1	17	25	$980	$210	13	2.76	**19**	▽
LINCOLN TOWN CAR									
4.6L 8, auto	LEV	17	24	$910	$180	14	2.65	**20**	○
4.6L 8, auto	TLEV	17	24	$910	$200	14	2.78	**19**	▽
4.6L 8, auto	Tier 1	17	24	$910	$210	14	2.81	**19**	▽
MERCEDES S320									
5.0L 6, auto [P]	LEV	17	24	$980	$180	14	2.67	**20**	○
5.0L 6, auto [P]	Tier 1	17	24	$980	$210	14	2.83	**18**	▽
MERCEDES S420									
4.2L 8, auto [P]	LEV	15	22	$1,080	$190	15	2.88	**18**	▽
4.2L 8, auto [P]	TLEV	15	22	$1,080	$220	15	3.04	**16**	✖
4.2L 8, auto [P]	Tier 1	15	22	$1,080	$220	15	3.07	**16**	✖
MERCEDES S500									
5.0L 8, auto [P]	LEV	15	21	$1,150	$190	16	2.95	**17**	▽
5.0L 8, auto [P]	TLEV	15	21	$1,150	$220	16	3.11	**16**	✖
5.0L 8, auto [P]	Tier 1	15	21	$1,150	$230	16	3.14	**16**	✖
MERCEDES S600									
6.0L 12, auto [P]	LEV	13	19	$1,300	$210	18	3.30	**14**	✖
6.0L 12, auto [P]	Tier 1	13	19	$1,300	$240	18	3.46	**13**	✖

[10] *CNG fuel economy is given as gasoline-equivalent MPG.*

LARGE CARS (cont.)

	Emission Standard	Fuel Economy City	Fuel Economy Hwy	Fuel Cost/yr	Health Cost/yr	GHG tons/yr	EDX ¢/mi	Green Score	Class Ranking
MERCURY GRAND MARQUIS									
4.6L 8, auto	LEV	17	24	$910	$180	14	2.65	**20**	○
4.6L 8, auto	Tier 1	17	24	$910	$210	14	2.81	**19**	▽
OLDSMOBILE EIGHTY EIGHT/REGENCY									
3.8L 6, auto	Tier 1	19	29	$750	$190	12	2.51	**22**	▲
3.8L 6, auto [P]	Tier 1	18	27	$930	$200	13	2.66	**20**	○
PONTIAC BONNEVILLE									
3.8L 6, auto	TLEV	19	28	$790	$190	12	2.58	**21**	○
3.8L 6, auto	Tier 1	19	28	$790	$200	12	2.61	**21**	○
3.8L 6, auto [P]	TLEV	18	27	$930	$200	13	2.66	**20**	○
3.8L 6, auto [P]	Tier 1	18	27	$930	$200	13	2.69	**20**	○
ROLLS ROYCE SILVER SPUR PARK WARD									
6.8L 8, auto [P]	Tier 1	10	15	$1,620	$280	22	4.16	**9**	✖
TOYOTA AVALON									
3.0L 6, auto	LEV	21	29	$720	$160	11	2.27	**25**	✔
3.0L 6, auto	Tier 1	21	29	$720	$190	11	2.42	**23**	▲

MINIVANS

	Emission Standard	Fuel Economy City	Fuel Economy Hwy	Fuel Cost/yr	Health Cost/yr	GHG tons/yr	EDX ¢/mi	Green Score	Class Ranking
CHEVROLET ASTRO (CARGO)									
4.3L 6, auto 4wd	LEV	16	20	$960	$230	15	3.13	**16**	▽
4.3L 6, auto	TLEV	16	21	$960	$250	15	3.22	**15**	▽
4.3L 6, auto	Tier 1	16	21	$960	$250	15	3.25	**15**	▽
4.3L 6, auto 4wd	Tier 1	16	20	$960	$260	15	3.34	**14**	✖
CHEVROLET ASTRO (PASSENGER)									
4.3L 6, auto 4wd	LEV	15	19	$1,010	$240	16	3.27	**15**	✖
4.3L 6, auto	TLEV	16	20	$960	$260	15	3.31	**14**	✖
4.3L 6, auto	Tier 1	16	20	$960	$260	15	3.34	**14**	✖
4.3L 6, auto 4wd	Tier 1	15	19	$1,010	$270	16	3.48	**13**	✖
CHEVROLET VENTURE									
3.3L 6, auto	TLEV	18	25	$860	$240	14	3.02	**17**	○
3.3L 6, auto	Tier 1	18	25	$860	$240	14	3.05	**16**	○
CHRYSLER EPIC									
Electric[11]	ZEV	1.5	1.4	$1,060	$130	10	1.94	**30**	✔

[11] *Electric vehicle fuel economy is given in mi/kWh.*

	Emission Standard	Fuel Economy City	Hwy	Fuel Cost/yr	Health Cost/yr	GHG tons/yr	EDX ¢/mi	Green Score	Class Ranking
MINIVANS (cont.)									
CHRYSLER TOWN & COUNTRY[12]									
3.3L 6, auto	LEV	18	24	$860	$210	14	2.85	**18**	▲
3.8L 6, auto	LEV	17	24	$910	$220	14	2.91	**18**	○
3.3L 6, auto	TLEV	18	24	$860	$240	14	3.03	**17**	○
3.3L 6, auto	TLEV	18	24	$860	$240	14	3.03	**17**	○
3.8L 6, auto Awd	LEV	16	23	$910	$230	15	3.05	**16**	○
3.3L 6, auto	Tier 1	18	24	$860	$250	14	3.06	**16**	○
3.3L 6, auto	Tier 1	18	24	$860	$250	14	3.06	**16**	○
3.3L 6, auto	TLEV	18	24	$860	$250	14	3.09	**16**	○
3.8L 6, auto	TLEV	17	24	$910	$240	14	3.09	**16**	○
3.3L 6, auto	Tier 1	18	24	$860	$250	14	3.12	**16**	▽
3.8L 6, auto	Tier 1	17	24	$910	$250	14	3.12	**16**	▽
3.8L 6, auto Awd	TLEV	16	23	$910	$250	15	3.23	**15**	▽
3.8L 6, auto Awd	Tier 1	16	23	$910	$260	15	3.26	**15**	✖
DODGE CARAVAN[13]									
2.4L 4, auto 3 sp	LEV	20	26	$790	$210	13	2.70	**20**	✔
3.0L 6, auto 4 sp	LEV	19	26	$820	$210	13	2.77	**19**	✔
3.3L 6, auto 4 sp	LEV	18	24	$860	$210	14	2.86	**18**	▲
2.4L 4, auto 3 sp	TLEV	20	26	$790	$230	13	2.88	**18**	▲
3.8L 6, auto 4 sp	LEV	17	24	$910	$220	14	2.91	**18**	○
2.4L 4, auto 3 sp	Tier 1	20	26	$790	$240	13	2.91	**18**	○
3.0L 6, auto 4 sp	TLEV	19	26	$820	$240	13	2.95	**17**	○
3.0L 6, auto 4 sp	Tier 1	19	26	$820	$240	13	2.98	**17**	○
3.8L 6, auto 4 sp	LEV	17	23	$910	$220	14	2.99	**17**	○
3.3L 6, auto 4 sp	TLEV	18	24	$860	$240	14	3.04	**16**	○
3.3L 6, auto 4 sp	Tier 1	18	24	$860	$250	14	3.07	**16**	○
3.8L 6, auto 4 sp	TLEV	17	24	$910	$240	14	3.09	**16**	○
3.8L 6, auto 4 sp	Tier 1	17	24	$910	$250	14	3.12	**16**	▽
3.8L 6, auto 4 sp	TLEV	17	23	$910	$250	14	3.17	**15**	▽
3.8L 6, auto 4 sp	Tier 1	17	23	$910	$250	15	3.20	**15**	▽
3.8L 6, auto Awd 4 sp	Tier 1	16	23	$910	$260	15	3.26	**15**	✖
FORD WINDSTAR VAN									
3.8L 6, auto	LEV	18	23	$860	$220	14	2.89	**18**	▲
3.0L 6, auto	LEV	17	23	$910	$220	14	2.93	**17**	○
FORD WINDSTAR WAGON									
3.0L 6, auto	LEV	17	23	$910	$220	14	2.93	**17**	○
3.8L 6, auto	LEV	17	23	$910	$220	15	3.00	**17**	○

[12/13] *An ethanol-gasoline flex-fuel version of this vehicle is also available.*

	Emission Standard	Fuel Economy City	Fuel Economy Hwy	Fuel Cost/yr	Health Cost/yr	GHG tons/yr	EDX ¢/mi	Green Score	Class Ranking
MINIVANS (cont.)									
GMC SAFARI (CARGO)									
4.3L 6, auto	TLEV	16	21	$960	$250	15	3.22	**15**	▽
4.3L 6, auto	Tier 1	16	21	$960	$250	15	3.25	**15**	▽
4.3L 6, auto Awd	TLEV	16	20	$960	$260	15	3.31	**14**	✖
4.3L 6, auto Awd	Tier 1	16	20	$960	$260	15	3.34	**14**	✖
GMC SAFARI (PASSENGER)									
4.3L 6, auto	TLEV	16	20	$960	$260	15	3.31	**14**	✖
4.3L 6, auto	Tier 1	16	20	$960	$260	15	3.34	**14**	✖
4.3L 6, auto Awd	TLEV	15	19	$1,010	$270	16	3.45	**13**	✖
4.3L 6, auto Awd	Tier 1	15	19	$1,010	$270	16	3.48	**13**	✖
HONDA ODYSSEY									
3.5L 6, auto	LEV	18	26	$820	$220	14	2.87	**18**	▲
3.5L 6, auto	Tier 1	18	26	$820	$250	14	3.08	**16**	○
MERCURY VILLAGER WAGON									
3.3L 6, auto	TLEV	17	24	$860	$240	14	3.07	**16**	○
3.3L 6, auto	Tier 1	17	24	$860	$250	14	3.10	**16**	▽
NISSAN QUEST									
3.3L 6, auto	LEV	17	24	$860	$220	14	2.88	**18**	▲
3.3L 6, auto	Tier 1	17	24	$860	$250	14	3.09	**16**	○
OLDSMOBILE SILHOUETTE									
3.3L 6, auto	TLEV	18	25	$860	$240	14	3.02	**17**	○
3.3L 6, auto	Tier 1	18	25	$860	$240	14	3.05	**16**	○
PLYMOUTH VOYAGER[14]									
2.4L 4, auto 3 sp	LEV	20	26	$790	$210	13	2.70	**20**	✔
3.0L 6, auto 4 sp	LEV	19	26	$820	$210	13	2.77	**19**	✔
3.3L 6, auto 4 sp	LEV	18	24	$860	$210	14	2.86	**18**	▲
2.4L 4, auto 3 sp	TLEV	20	26	$790	$230	13	2.88	**18**	▲
3.8L 6, auto 4 sp	LEV	17	24	$910	$220	14	2.90	**18**	○
2.4L 4, auto 3 sp	Tier 1	20	26	$790	$240	13	2.91	**18**	○
3.0L 6, auto 4 sp	TLEV	19	26	$820	$240	13	2.95	**17**	○
3.0L 6, auto 4 sp	Tier 1	19	26	$820	$240	13	2.98	**17**	○
3.3L 6, auto 4 sp	TLEV	18	24	$860	$240	14	3.04	**16**	○
3.3L 6, auto 4 sp	Tier 1	18	24	$860	$250	14	3.07	**16**	○
3.8L 6, auto 4 sp	TLEV	17	24	$910	$240	14	3.08	**16**	○
3.8L 6, auto 4 sp	Tier 1	17	24	$910	$250	14	3.11	**16**	▽

[14] *An ethanol-gasoline flex-fuel version of this vehicle is also available.*

	Emission Standard	Fuel Economy City	Hwy	Fuel Cost/yr	Health Cost/yr	GHG tons/yr	EDX ¢/mi	Green Score	Class Ranking
MINIVANS (cont.)									
PONTIAC TRANS SPORT									
3.3L 6, auto	TLEV	18	25	$860	$240	14	3.02	**17**	○
3.3L 6, auto	Tier 1	18	25	$860	$240	14	3.05	**16**	○
TOYOTA SIENNA									
3.0L 6, auto	Tier 1	18	24	$820	$240	13	3.02	**17**	○
LARGE VANS									
CHEVROLET G1500/G2500									
4.3L 6, auto	Tier 1	15	19	$1,080	$270	17	3.53	**13**	○
5.0L 8, auto	Tier 1	14	18	$1,080	$280	17	3.59	**12**	○
5.7L 8, auto	Tier 1	14	18	$1,150	$280	18	3.70	**12**	○
5.0L 8, auto	Tier 1	14	17	$1,150	$280	18	3.74	**12**	▽
5.0L 8, auto	HDT	14	17	$1,150	$360	18	4.26	**9**	✖
DODGE B1500 VAN									
3.9L 6, auto 3 sp	LEV	15	17	$1,080	$240	17	3.33	**14**	▲
5.2L 8, auto 4 sp	LEV	13	19	$1,150	$240	17	3.40	**14**	▲
3.9L 6, auto 3 sp	Tier 1	15	17	$1,080	$270	17	3.54	**13**	○
5.2L 8, auto 4 sp	Tier 1	13	19	$1,150	$270	17	3.61	**12**	○
5.9L 8, auto 4 sp	LEV	13	17	$1,230	$250	18	3.62	**12**	○
5.9L 8, auto 4 sp	Tier 1	13	17	$1,230	$290	19	3.83	**11**	✖
DODGE B1500 WAGON									
5.2L 8, auto 4 sp	Tier 1	13	19	$1,150	$280	17	3.66	**12**	○
3.9L 6, auto 3 sp	Tier 1	15	16	$1,150	$280	18	3.68	**12**	○
DODGE B2500 VAN									
5.2L 8, auto [CNG][15]	SULEV	14	19	$660	$100	14	2.16	**26**	✔
5.2L 8, auto	LEV	13	19	$1,150	$250	17	3.45	**13**	▲
5.9L 8, auto	LEV	13	17	$1,230	$250	18	3.62	**12**	○
5.2L 8, auto	Tier 1	13	19	$1,150	$280	17	3.66	**12**	○
5.9L 8, auto	Tier 1	13	17	$1,230	$290	19	3.83	**11**	✖
DODGE B2500 WAGON									
5.2L 8, auto [CNG][16]	SULEV	14	19	$680	$110	15	2.21	**26**	✔
5.2L 8, auto	LEV	13	18	$1,150	$260	18	3.57	**13**	○
5.9L 8, auto	LEV	12	17	$1,230	$260	19	3.74	**12**	▽
5.2L 8, auto	Tier 1	13	18	$1,150	$290	18	3.78	**11**	▽

[15/16] *CNG fuel economy is given as gasoline-equivalent MPG.*

	Emission Standard	Fuel Economy City	Hwy	Fuel Cost/yr	Health Cost/yr	GHG tons/yr	EDX ¢/mi	Green Score	Class Ranking
LARGE VANS (cont.)									
DODGE B2500 WAGON (cont.)									
5.9L 8, auto	Tier 1	12	17	$1,230	$290	19	3.95	**10**	✖
5.9L 8, auto	HDT	12	17	$1,260	$370	20	4.52	**8**	✖
FORD E150 CLUB WAGON									
4.2L 6, auto	Tier 1	14	18	$1,150	$290	18	3.75	**12**	▽
4.6L 8, auto	Tier 1	13	17	$1,150	$290	18	3.79	**11**	▽
5.4L 8, auto	Tier 1	12	17	$1,230	$360	20	4.43	**8**	✖
FORD E150 ECONOLINE									
4.2L 6, auto	LEV	14	18	$1,080	$250	17	3.43	**14**	▲
4.6L 8, auto	LEV	13	18	$1,150	$250	18	3.50	**13**	▲
5.4L 8, auto	LEV	13	17	$1,150	$250	18	3.51	**13**	○
4.2L 6, auto	Tier 1	14	18	$1,080	$280	17	3.64	**12**	○
4.6L 8, auto	Tier 1	13	18	$1,150	$280	18	3.71	**12**	▽
5.4L 8, auto	Tier 1	13	17	$1,150	$280	18	3.72	**12**	▽
FORD E250 ECONOLINE[17]									
5.4L 8, auto [CNG][18]	SULEV	12	15	$870	$130	17	2.64	**20**	✔
5.4L 8, auto [CNG][19]	ULEV	12	15	$870	$170	18	2.96	**17**	▲
4.2L 6, auto	LEV	14	18	$1,150	$250	18	3.47	**13**	▲
5.4L 8, auto	LEV	13	18	$1,230	$260	19	3.66	**12**	○
4.2L 6, auto	Tier 1	14	18	$1,150	$280	18	3.69	**12**	○
5.4L 8, auto	Tier 1	13	18	$1,230	$290	19	3.87	**11**	✖
5.4L 8, auto	HDT	12	17	$1,220	$360	19	4.43	**8**	✖
GMC G1500/G2500 SAVANA									
4.3L 6, auto	Tier 1	15	19	$1,080	$270	17	3.53	**13**	○
5.0L 8, auto	Tier 1	14	18	$1,080	$280	17	3.59	**12**	○
5.7L 8, auto	Tier 1	14	18	$1,150	$280	18	3.70	**12**	○
5.0L 8, auto	Tier 1	14	17	$1,150	$280	18	3.74	**12**	▽
5.7L 8, auto	HDT	14	17	$1,160	$360	18	4.26	**9**	✖
VOLKSWAGEN EUROVAN									
2.8L 6, auto	Tier 1	15	20	$1,010	$270	16	3.44	**13**	▲
VOLKSWAGEN EUROVAN CAMPER									
2.8L 6, auto	Tier 1	14	19	$1,080	$280	17	3.67	**12**	○

[17] *CNG fuel economy is given as gasoline-equivalent MPG.*

[18/19] *A propane-gasoline bi-fuel version of this vehicle is also available.*

	Emission Standard	Fuel Economy City	Fuel Economy Hwy	Fuel Cost/yr	Health Cost/yr	GHG tons/yr	EDX ¢/mi	Green Score	Class Ranking
COMPACT PICKUPS									
CHEVROLET S10 PICKUP									
Electric[20]	ZEV	2.2	2.4	$640	$80	6	1.19	**46**	✔
2.2L 4, manual	LEV	23	29	$690	$160	11	2.22	**26**	▲
2.2L 4, manual	TLEV	23	29	$690	$180	11	2.35	**24**	▲
2.2L 4, manual	Tier 1	23	29	$690	$180	11	2.38	**23**	▲
2.2L 4, auto	LEV	19	26	$790	$170	12	2.44	**23**	▲
2.2L 4, auto	TLEV	19	26	$790	$190	12	2.57	**21**	○
2.2L 4, auto	Tier 1	19	26	$790	$200	12	2.60	**21**	○
4.3L 6, manual	LEV	17	23	$910	$180	14	2.62	**21**	○
4.3L 6, auto	LEV	17	22	$960	$180	14	2.70	**20**	○
4.3L 6, manual	TLEV	17	23	$910	$200	14	2.75	**19**	○
4.3L 6, manual	Tier 1	17	23	$910	$200	14	2.78	**19**	○
4.3L 6, auto	TLEV	17	22	$960	$200	14	2.83	**18**	○
4.3L 6, auto	Tier 1	17	22	$960	$210	14	2.86	**18**	○
4.3L 6, manual	LEV	17	23	$910	$220	14	2.92	**18**	○
4.3L 6, auto	LEV	17	22	$960	$220	15	3.00	**17**	○
4.3L 6, manual 4wd	LEV	17	21	$960	$220	15	3.00	**17**	○
4.3L 6, auto 4wd	LEV	16	21	$960	$220	15	3.05	**16**	○
4.3L 6, manual	TLEV	17	23	$910	$240	14	3.10	**16**	○
4.3L 6, manual	Tier 1	17	23	$910	$250	14	3.13	**16**	○
4.3L 6, auto	TLEV	17	22	$960	$250	15	3.18	**15**	○
4.3L 6, manual 4wd	TLEV	17	21	$960	$250	15	3.18	**15**	○
4.3L 6, auto	Tier 1	17	22	$960	$250	15	3.21	**15**	▽
4.3L 6, manual 4wd	Tier 1	17	21	$960	$250	15	3.21	**15**	▽
4.3L 6, auto 4wd	TLEV	16	21	$960	$250	15	3.23	**15**	▽
4.3L 6, auto 4wd	Tier 1	16	21	$960	$250	15	3.26	**15**	▽
DODGE DAKOTA PICKUP									
2.5L 4, manual	TLEV	20	24	$790	$190	12	2.57	**21**	○
2.5L 4, manual	Tier 1	20	24	$790	$200	12	2.60	**21**	○
2.5L 4, manual	TLEV	20	24	$790	$240	13	2.92	**18**	○
2.5L 4, manual	Tier 1	20	24	$790	$240	13	2.94	**17**	○
3.9L 6, manual	TLEV	16	22	$960	$250	15	3.20	**15**	▽
3.9L 6, manual	Tier 1	16	22	$960	$250	15	3.23	**15**	▽
3.9L 6, auto	TLEV	16	21	$960	$250	15	3.24	**15**	▽
3.9L 6, auto	Tier 1	16	21	$960	$260	15	3.28	**15**	▽
3.9L 6, manual 4wd	Tier 1	15	19	$1,010	$260	16	3.38	**14**	✖
3.9L 6, manual 4wd	TLEV	15	19	$1,010	$260	16	3.40	**14**	✖
3.9L 6, manual 4wd	Tier 1	15	19	$1,010	$270	16	3.43	**13**	✖

[20] *Electric vehicle fuel economy is given in mi/kWh.*

	Emission Standard	Fuel Economy City	Fuel Economy Hwy	Fuel Cost/yr	Health Cost/yr	GHG tons/yr	EDX ¢/mi	Green Score	Class Ranking
COMPACT PICKUPS (cont.)									
DODGE DAKOTA PICKUP (cont.)									
3.9L 6, auto 4wd	TLEV	14	18	$1,080	$270	17	3.52	**13**	✖
3.9L 6, auto 4wd	Tier 1	14	18	$1,080	$270	17	3.55	**13**	✖
5.2L 8, auto	TLEV	14	18	$1,150	$270	17	3.56	**13**	✖
5.2L 8, auto	Tier 1	14	18	$1,150	$270	17	3.59	**12**	✖
5.2L 8, manual	TLEV	13	18	$1,150	$270	17	3.59	**12**	✖
5.2L 8, manual	Tier 1	13	18	$1,150	$270	17	3.63	**12**	✖
5.2L 8, manual 4wd	TLEV	13	18	$1,150	$270	18	3.64	**12**	✖
5.2L 8, manual 4wd	Tier 1	13	18	$1,150	$280	18	3.67	**12**	✖
5.2L 8, auto 4wd	TLEV	13	16	$1,230	$280	18	3.73	**12**	✖
5.2L 8, auto 4wd	Tier 1	13	16	$1,230	$280	18	3.76	**11**	✖
5.9L 8, auto	TLEV	12	17	$1,230	$280	19	3.80	**11**	✖
5.9L 8, auto	Tier 1	12	17	$1,230	$280	19	3.83	**11**	✖
FORD RANGER PICKUP									
Electric[21]	ZEV	2.6	2.3	$620	$100	7	1.36	**42**	✔
2.5L 4, manual	LEV	22	27	$720	$170	12	2.31	**24**	▲
2.5L 4, auto 4 sp	LEV	20	25	$790	$170	12	2.42	**23**	▲
2.5L 4, manual	TLEV	22	27	$720	$190	12	2.44	**23**	▲
2.5L 4, manual	Tier 1	22	27	$720	$190	12	2.47	**22**	▲
2.5L 4, auto 4 sp	TLEV	20	25	$790	$190	12	2.55	**21**	○
2.5L 4, auto 4 sp	Tier 1	20	25	$790	$190	12	2.58	**21**	○
4.0L 6, manual	LEV	17	22	$910	$180	14	2.63	**20**	○
4.0L 6, manual 4wd	LEV	17	20	$960	$190	14	2.73	**19**	○
3.0L 6, manual FFV	Tier 1	18	23	$860	$200	13	2.74	**19**	○
4.0L 6, auto 5 sp	LEV	16	21	$960	$190	15	2.75	**19**	○
4.0L 6, manual	TLEV	17	22	$910	$200	14	2.76	**19**	○
4.0L 6, manual	Tier 1	17	22	$910	$200	14	2.79	**19**	○
3.0L 6, manual 4wd FFV	Tier 1	18	21	$910	$200	14	2.80	**19**	○
3.0L 6, auto FFV 4 sp	Tier 1	17	22	$910	$210	14	2.84	**18**	○
4.0L 6, manual 4wd	TLEV	17	20	$960	$200	14	2.86	**18**	○
4.0L 6, auto 5 sp	TLEV	16	21	$960	$210	15	2.88	**18**	○
4.0L 6, manual 4wd	Tier 1	17	20	$960	$210	14	2.89	**18**	○
4.0L 6, auto 5 sp	Tier 1	16	21	$960	$210	15	2.91	**18**	○
4.0L 6, manual	LEV	17	22	$910	$220	14	2.92	**18**	○
4.0L 6, manual 4wd	LEV	17	20	$960	$220	15	3.03	**17**	○
4.0L 6, auto 5 sp	LEV	16	21	$960	$220	15	3.05	**16**	○
4.0L 6, auto 4wd 5 sp	LEV	16	20	$960	$230	15	3.09	**16**	○
4.0L 6, manual	TLEV	17	22	$910	$240	14	3.10	**16**	○

[21] *Electric vehicle fuel economy is given in mi/kWh.*

	Emission Standard	Fuel Economy City	Hwy	Fuel Cost/yr	Health Cost/yr	GHG tons/yr	EDX ¢/mi	Green Score	Class Ranking
COMPACT PICKUPS (cont.)									
FORD RANGER PICKUP (cont.)									
4.0L 6, manual	Tier 1	17	22	$910	$250	14	3.13	**16**	○
3.0L 6, manual 4wd FFV	Tier 1	18	21	$910	$250	14	3.14	**16**	○
3.0L 6, auto FFV 4 sp	Tier 1	17	22	$910	$250	15	3.19	**15**	○
4.0L 6, manual 4wd	TLEV	17	20	$960	$250	15	3.21	**15**	▽
4.0L 6, auto 5 sp	TLEV	16	21	$960	$250	15	3.23	**15**	▽
4.0L 6, manual 4wd	Tier 1	17	20	$960	$250	15	3.24	**15**	▽
4.0L 6, auto 5 sp	Tier 1	16	21	$960	$250	15	3.26	**15**	▽
4.0L 6, auto 4wd 5 sp	TLEV	16	20	$960	$250	15	3.27	**15**	▽
4.0L 6, auto 4wd 5 sp	Tier 1	16	20	$960	$260	15	3.30	**14**	✖
3.0L 6, auto 4wd FFV 4sp	Tier 1	15	19	$1,010	$260	16	3.40	**14**	✖
GMC SONOMA									
2.2L 4, manual	LEV	23	29	$690	$160	11	2.22	**26**	▲
2.2L 4, manual	TLEV	23	29	$690	$180	11	2.35	**24**	▲
2.2L 4, manual	Tier 1	23	29	$690	$180	11	2.38	**23**	▲
2.2L 4, auto	LEV	19	26	$790	$170	12	2.44	**23**	▲
2.2L 4, auto	TLEV	19	26	$790	$190	12	2.57	**21**	○
2.2L 4, auto	Tier 1	19	26	$790	$200	12	2.60	**21**	○
4.3L 6, manual	LEV	17	23	$910	$180	14	2.62	**21**	○
4.3L 6, auto	LEV	17	22	$960	$180	14	2.70	**20**	○
4.3L 6, manual	TLEV	17	23	$910	$200	14	2.75	**19**	○
4.3L 6, manual	Tier 1	17	23	$910	$200	14	2.78	**19**	○
4.3L 6, auto	TLEV	17	22	$960	$200	14	2.83	**18**	○
4.3L 6, auto	Tier 1	17	22	$960	$210	14	2.86	**18**	○
4.3L 6, manual	LEV	17	23	$910	$220	14	2.92	**18**	○
4.3L 6, auto	LEV	17	22	$960	$220	15	3.00	**17**	○
4.3L 6, manual 4wd	LEV	17	21	$960	$220	15	3.00	**17**	○
4.3L 6, auto 4wd	LEV	16	21	$960	$220	15	3.05	**16**	○
4.3L 6, manual	TLEV	17	23	$910	$240	14	3.10	**16**	○
4.3L 6, manual	Tier 1	17	23	$910	$250	14	3.13	**16**	○
4.3L 6, auto	TLEV	17	22	$960	$250	15	3.18	**15**	○
4.3L 6, manual 4wd	TLEV	17	21	$960	$250	15	3.18	**15**	○
4.3L 6, auto	Tier 1	17	22	$960	$250	15	3.21	**15**	▽
4.3L 6, manual 4wd	Tier 1	17	21	$960	$250	15	3.21	**15**	▽
4.3L 6, auto 4wd	TLEV	16	21	$960	$250	15	3.23	**15**	▽
4.3L 6, auto 4wd	Tier 1	16	21	$960	$250	15	3.26	**15**	▽

See pages 32-33 for explanation of data and symbols.

	Emission Standard	Fuel Economy City	Hwy	Fuel Cost/yr	Health Cost/yr	GHG tons/yr	EDX ¢/mi	Green Score	Class Ranking
COMPACT PICKUPS (cont.)									
ISUZU HOMBRE PICKUP									
2.2L 4, manual	LEV	23	29	$690	$160	11	2.22	**26**	▲
2.2L 4, manual	TLEV	23	29	$690	$180	11	2.35	**24**	▲
2.2L 4, manual	Tier 1	23	29	$690	$180	11	2.38	**23**	▲
2.2L 4, auto	LEV	19	26	$790	$170	12	2.44	**23**	▲
2.2L 4, auto	TLEV	19	26	$790	$190	12	2.57	**21**	○
2.2L 4, auto	Tier 1	19	26	$790	$200	12	2.60	**21**	○
4.3L 6, manual	LEV	17	23	$910	$220	14	2.92	**18**	○
4.3L 6, manual 4wd	LEV	17	21	$960	$220	15	3.00	**17**	○
4.3L 6, auto	LEV	16	21	$960	$220	15	3.04	**17**	○
4.3L 6, auto 4wd	LEV	16	21	$960	$220	15	3.05	**16**	○
4.3L 6, manual	TLEV	17	23	$910	$240	14	3.10	**16**	○
4.3L 6, manual	Tier 1	17	23	$910	$250	14	3.13	**16**	○
4.3L 6, manual 4wd	TLEV	17	21	$960	$250	15	3.18	**15**	○
4.3L 6, manual 4wd	Tier 1	17	21	$960	$250	15	3.21	**15**	▽
4.3L 6, auto	TLEV	16	21	$960	$250	15	3.22	**15**	▽
4.3L 6, auto 4wd	TLEV	16	21	$960	$250	15	3.23	**15**	▽
4.3L 6, auto	Tier 1	16	21	$960	$250	15	3.25	**15**	▽
4.3L 6, auto 4wd	Tier 1	16	21	$960	$250	15	3.26	**15**	▽
MAZDA B2500									
2.5L 4, manual	LEV	22	27	$720	$170	12	2.31	**24**	▲
2.5L 4, auto	LEV	20	25	$790	$170	12	2.42	**23**	▲
2.5L 4, manual	TLEV	22	27	$720	$190	12	2.44	**23**	▲
2.5L 4, manual	Tier 1	22	27	$720	$190	12	2.47	**22**	▲
2.5L 4, auto	TLEV	20	25	$790	$190	12	2.55	**21**	○
2.5L 4, auto	Tier 1	20	25	$790	$190	12	2.58	**21**	○
MAZDA B3000 FFV									
3.0L 6, manual FFV	LEV	18	23	$860	$180	13	2.58	**21**	○
3.0L 6, manual 4wd FFV	LEV	18	21	$910	$180	14	2.64	**20**	○
3.0L 6, auto FFV	LEV	17	22	$910	$180	14	2.68	**20**	○
3.0L 6, manual FFV	TLEV	18	23	$860	$200	13	2.71	**20**	○
3.0L 6, manual FFV	Tier 1	18	23	$860	$200	13	2.74	**19**	○
3.0L 6, manual 4wd FFV	TLEV	18	21	$910	$200	14	2.77	**19**	○
3.0L 6, manual 4wd FFV	Tier 1	18	21	$910	$200	14	2.80	**19**	○
3.0L 6, auto FFV	TLEV	17	22	$910	$200	14	2.81	**19**	○
3.0L 6, auto FFV	Tier 1	17	22	$910	$210	14	2.84	**18**	○
3.0L 6, manual 4wd FFV	LEV	18	21	$910	$220	14	2.93	**17**	○

See pages 32-33 for explanation of data and symbols.

	Emission Standard	Fuel Economy City	Fuel Economy Hwy	Fuel Cost/yr	Health Cost/yr	GHG tons/yr	EDX ¢/mi	Green Score	Class Ranking
COMPACT PICKUPS (cont.)									
MAZDA B3000 FFV (cont.)									
3.0L 6, manual 4wd FFV	TLEV	18	21	$910	$240	14	3.11	**16**	○
3.0L 6, manual 4wd FFV	Tier 1	18	21	$910	$250	14	3.14	**16**	○
3.0L 6, auto 4wd FFV	LEV	15	19	$1,010	$230	16	3.19	**15**	○
3.0L 6, auto 4wd FFV	TLEV	15	19	$1,010	$260	16	3.37	**14**	✖
3.0L 6, auto 4wd FFV	Tier 1	15	19	$1,010	$260	16	3.40	**14**	✖
NISSAN FRONTIER TRUCK									
2.4L 4, manual	LEV	22	26	$720	$160	11	2.23	**25**	▲
2.4L 4, manual	TLEV	22	26	$720	$180	11	2.36	**24**	▲
2.4L 4, manual	Tier 1	22	26	$720	$180	11	2.39	**23**	▲
2.4L 4, auto	LEV	20	24	$820	$170	13	2.46	**22**	▲
2.4L 4, auto	TLEV	20	24	$820	$190	13	2.59	**21**	○
2.4L 4, auto	Tier 1	20	24	$820	$200	13	2.62	**21**	○
2.4L 4, manual 4wd	LEV	18	21	$910	$220	14	2.95	**17**	○
3.3L 6, manual 4wd	LEV	16	19	$1,010	$230	15	3.11	**16**	○
2.4L 4, manual 4wd	TLEV	18	21	$910	$250	14	3.13	**16**	○
2.4L 4, manual 4wd	Tier 1	18	21	$910	$250	14	3.16	**16**	○
3.3L 6, auto 4wd	LEV	15	19	$1,010	$230	16	3.18	**15**	○
3.3L 6, manual 4wd	TLEV	16	19	$1,010	$250	15	3.29	**15**	▽
3.3L 6, manual 4wd	Tier 1	16	19	$1,010	$260	15	3.32	**14**	✖
3.3L 6, auto 4wd	TLEV	15	19	$1,010	$260	16	3.36	**14**	✖
3.3L 6, auto 4wd	Tier 1	15	19	$1,010	$260	16	3.39	**14**	✖
TOYOTA TACOMA									
2.4L 4, manual	Tier 1	22	27	$720	$180	11	2.39	**23**	▲
2.4L 4, auto	Tier 1	21	24	$790	$190	12	2.48	**22**	▲
3.4L 6, manual	Tier 1	19	24	$820	$200	13	2.63	**20**	○
3.4L 6, auto	Tier 1	19	23	$860	$200	13	2.68	**20**	○
2.7L 4, auto	Tier 1	19	21	$860	$200	13	2.71	**20**	○
2.7L 4, auto 4wd	Tier 1	18	21	$910	$200	14	2.78	**19**	○
2.7L 4, manual 4wd	Tier 1	17	21	$910	$210	14	2.82	**18**	○
3.4L 6, auto	Tier 1	19	23	$860	$240	14	3.03	**17**	○
2.7L 4, auto 4wd	Tier 1	18	21	$910	$250	14	3.13	**16**	○
3.4L 6, auto 4wd	Tier 1	17	20	$960	$250	15	3.22	**15**	▽
3.4L 6, manual 4wd	Tier 1	17	19	$960	$250	15	3.26	**15**	▽

See pages 32-33 for explanation of data and symbols.

	Emission Standard	Fuel Economy City	Hwy	Fuel Cost/yr	Health Cost/yr	GHG tons/yr	EDX ¢/mi	Green Score	Class Ranking
STANDARD PICKUPS									
CHEVROLET C1500 PICKUP									
5.0L 8, auto	Tier 1	15	19	$1,080	$270	17	3.52	**13**	○
5.7L 8, auto	Tier 1	14	18	$1,080	$280	17	3.63	**12**	▽
CHEVROLET C1500 SILVERADO									
4.3L 6, manual	LEV	17	23	$910	$220	14	2.94	**17**	▲
4.3L 6, auto	LEV	16	20	$960	$220	15	3.07	**16**	▲
4.8L 8, manual	LEV	16	20	$1,010	$230	15	3.10	**16**	▲
4.8L 8, auto	LEV	16	21	$960	$230	15	3.11	**16**	▲
4.3L 6, manual	Tier 1	17	23	$910	$250	14	3.15	**16**	▲
4.8L 8, manual	LEV	16	20	$1,010	$230	16	3.16	**16**	▲
4.3L 6, auto	Tier 1	16	20	$960	$260	15	3.28	**15**	▲
4.8L 8, manual	Tier 1	16	20	$1,010	$260	15	3.31	**14**	○
4.8L 8, auto	Tier 1	16	21	$960	$260	15	3.32	**14**	○
5.3L 8, auto	Tier 1	16	20	$1,010	$260	16	3.38	**14**	○
CHEVROLET C2500 PICKUP									
5.7L 8, auto	HDT	14	18	$1,120	$350	18	4.21	**9**	✖
CHEVROLET C2500 SILVERADO									
5.3L 8, auto	LEV	15	19	$1,010	$240	16	3.28	**15**	▲
5.3L 8, auto	Tier 1	15	19	$1,010	$270	16	3.49	**13**	○
6.0L 8, auto	LEV	12	16	$1,330	$260	19	3.76	**11**	▽
6.0L 8, auto	Tier 1	12	16	$1,330	$290	20	3.97	**10**	✖
6.0L 8, auto	HDT	12	16	$1,310	$370	20	4.53	**8**	✖
CHEVROLET K1500 PICKUP									
5.0L 8, auto	Tier 1	14	18	$1,080	$280	17	3.65	**12**	▽
5.7L 8, auto	Tier 1	13	17	$1,150	$290	18	3.80	**11**	✖
CHEVROLET K1500 SILVERADO									
4.3L 6, auto	LEV	16	20	$960	$230	15	3.13	**16**	▲
4.8L 8, manual	LEV	15	19	$1,080	$240	16	3.26	**15**	▲
4.8L 8, auto	LEV	15	18	$1,080	$240	16	3.27	**15**	▲
5.3L 8, auto	LEV	15	18	$1,080	$240	17	3.33	**14**	○
4.3L 6, auto	Tier 1	16	20	$960	$260	15	3.34	**14**	○
4.3L 6, manual	LEV	15	18	$1,080	$240	17	3.35	**14**	○
4.8L 8, manual	Tier 1	15	19	$1,080	$270	16	3.48	**13**	○
4.8L 8, auto	Tier 1	15	18	$1,080	$270	16	3.49	**13**	○
5.3L 8, auto	Tier 1	15	18	$1,080	$270	17	3.54	**13**	○
4.3L 6, manual	Tier 1	15	18	$1,080	$270	17	3.56	**13**	○
4.8L 8, auto	HDT	15	18	$1,070	$350	17	4.09	**10**	✖

See pages 32-33 for explanation of data and symbols.

STANDARD PICKUPS (cont.)

	Emission Standard	Fuel Economy City	Fuel Economy Hwy	Fuel Cost/yr	Health Cost/yr	GHG tons/yr	EDX ¢/mi	Green Score	Class Ranking
DODGE DURANGO									
3.9L 6, auto 4wd	LEV	14	18	$1,080	$240	17	3.34	**14**	○
3.9L 6, auto	Tier 1	15	20	$1,010	$270	16	3.46	**13**	○
3.9L 6, auto 4wd	Tier 1	14	18	$1,080	$270	17	3.55	**13**	○
5.2L 8, auto	Tier 1	14	19	$1,080	$270	17	3.60	**12**	○
5.2L 8, auto 4wd	LEV	13	17	$1,230	$250	18	3.60	**12**	▽
5.9L 8, auto 4wd	LEV	12	16	$1,230	$260	19	3.76	**11**	▽
5.9L 8, auto	Tier 1	12	17	$1,230	$280	19	3.81	**11**	✖
5.2L 8, auto 4wd	Tier 1	13	17	$1,230	$290	18	3.81	**11**	✖
5.9L 8, auto 4wd	Tier 1	12	16	$1,230	$290	20	3.97	**10**	✖
DODGE RAM 1500 PICKUP									
3.9L 6, manual	LEV	15	21	$1,010	$230	16	3.17	**15**	▲
3.9L 6, auto	LEV	15	20	$1,010	$240	16	3.25	**15**	▲
3.9L 6, manual	Tier 1	15	21	$1,010	$260	16	3.39	**14**	○
5.2L 8, auto	LEV	13	19	$1,150	$240	17	3.40	**14**	○
5.2L 8, manual	LEV	13	18	$1,150	$240	17	3.43	**14**	○
3.9L 6, auto	Tier 1	15	20	$1,010	$270	16	3.46	**13**	○
5.2L 8, auto	LEV	13	19	$1,150	$250	18	3.51	**13**	○
5.2L 8, manual	LEV	13	18	$1,150	$250	18	3.54	**13**	○
5.9L 8, auto	LEV	13	17	$1,230	$250	18	3.57	**13**	○
5.2L 8, manual 4wd	LEV	13	17	$1,230	$250	18	3.60	**12**	▽
5.2L 8, auto	Tier 1	13	19	$1,150	$270	17	3.61	**12**	▽
5.2L 8, manual	Tier 1	13	18	$1,150	$280	17	3.64	**12**	▽
5.9L 8, auto	LEV	13	17	$1,230	$260	19	3.68	**12**	▽
5.2L 8, auto 4wd	LEV	12	16	$1,230	$260	19	3.70	**12**	▽
5.2L 8, auto	Tier 1	13	19	$1,150	$280	18	3.72	**12**	▽
5.2L 8, manual	Tier 1	13	18	$1,150	$290	18	3.75	**12**	▽
5.9L 8, auto	Tier 1	13	17	$1,230	$280	18	3.79	**11**	▽
5.9L 8, auto 4wd	LEV	12	16	$1,330	$260	20	3.79	**11**	▽
5.2L 8, manual 4wd	Tier 1	13	17	$1,230	$290	18	3.81	**11**	✖
5.9L 8, auto	Tier 1	13	17	$1,230	$290	19	3.90	**11**	✖
5.2L 8, auto 4wd	Tier 1	12	16	$1,230	$290	19	3.91	**11**	✖
5.2L 8, manual 4wd	LEV	13	17	$1,230	$300	19	3.99	**10**	✖
5.9L 8, auto 4wd	Tier 1	12	16	$1,330	$290	20	4.01	**10**	✖
5.2L 8, auto 4wd	LEV	12	16	$1,230	$310	20	4.09	**10**	✖
5.9L 8, auto 4wd	LEV	12	16	$1,330	$310	20	4.18	**9**	✖
5.2L 8, manual 4wd	Tier 1	13	17	$1,230	$350	19	4.33	**9**	✖
5.2L 8, auto 4wd	Tier 1	12	16	$1,230	$360	20	4.43	**8**	✖
5.9L 8, auto 4wd	Tier 1	12	16	$1,330	$360	20	4.52	**8**	✖

See pages 32-33 for explanation of data and symbols.

	Emission Standard	Fuel Economy City	Hwy	Fuel Cost/yr	Health Cost/yr	GHG tons/yr	EDX ¢/mi	Green Score	Class Ranking
STANDARD PICKUPS (cont.)									
DODGE RAM 2500 PICKUP									
5.9L 8, auto	HDT	12	16	$1,320	$380	21	4.68	**7**	✖
FORD F150 PICKUP[22]									
4.2L 6, manual	Tier 1	16	21	$960	$250	15	3.23	**15**	▲
4.2L 6, auto	Tier 1	16	20	$1,010	$260	16	3.33	**14**	○
4.6L 8, auto	Tier 1	15	19	$1,080	$270	16	3.50	**13**	○
4.2L 6, auto 4wd	Tier 1	15	18	$1,080	$270	17	3.54	**13**	○
4.2L 6, manual 4wd	Tier 1	15	18	$1,080	$270	17	3.54	**13**	○
4.6L 8, manual	Tier 1	14	19	$1,080	$270	17	3.56	**13**	○
5.4L 8, auto	Tier 1	13	17	$1,150	$280	18	3.67	**12**	▽
4.6L 8, manual 4wd	Tier 1	14	17	$1,150	$280	18	3.74	**12**	▽
4.6L 8, auto 4wd	Tier 1	14	17	$1,150	$280	18	3.75	**12**	▽
5.4L 8, auto 4wd	Tier 1	13	16	$1,230	$290	19	3.90	**11**	✖
FORD F250 PICKUP[23]									
5.4L 8, auto [CNG][24]	SULEV	12	15	$870	$110	17	2.49	**22**	✔
5.4L 8, auto [CNG][25]	ULEV	12	15	$870	$140	17	2.71	**20**	✔
4.6L 8, manual	Tier 1	14	19	$1,080	$270	17	3.56	**13**	○
4.6L 8, auto	Tier 1	14	18	$1,150	$270	17	3.63	**12**	▽
5.4L 8, auto	Tier 1	13	17	$1,150	$280	18	3.72	**12**	▽
4.6L 8, auto 4wd	Tier 1	13	16	$1,150	$280	18	3.79	**11**	▽
4.6L 8, manual 4wd	Tier 1	13	17	$1,150	$280	18	3.80	**11**	✖
5.4L 8, auto 4wd	Tier 1	12	15	$1,330	$300	20	4.06	**10**	✖
5.4L 8, auto	HDT	13	17	$1,180	$360	19	4.31	**9**	✖
GMC C1500 SIERRA									
4.3L 6, manual	LEV	17	23	$910	$220	14	2.93	**17**	▲
4.3L 6, auto	LEV	16	20	$960	$220	15	3.07	**16**	▲
4.8L 8, manual	LEV	16	20	$1,010	$230	15	3.10	**16**	▲
4.8L 8, auto	LEV	16	21	$960	$230	15	3.11	**16**	▲
4.3L 6, manual	Tier 1	17	23	$910	$250	14	3.14	**16**	▲
5.3L 8, auto	LEV	16	20	$1,010	$230	16	3.17	**15**	▲
4.8L 8, manual	LEV	16	20	$1,010	$240	16	3.21	**15**	▲
4.3L 6, auto	Tier 1	16	20	$960	$260	15	3.28	**15**	▲
4.8L 8, manual	Tier 1	16	20	$1,010	$260	15	3.31	**14**	○
5.0L 8, auto	LEV	15	19	$1,080	$240	17	3.32	**14**	○
4.8L 8, auto	Tier 1	16	21	$960	$260	15	3.32	**14**	○
5.3L 8, auto	Tier 1	16	20	$1,010	$260	16	3.38	**14**	○

[22/23] *A propane-gasoline bi-fuel version of this vehicle is also available.*
[24/25] *CNG fuel economy is given as gasoline-equivalent MPG.*

	Emission Standard	Fuel Economy City	Fuel Economy Hwy	Fuel Cost/yr	Health Cost/yr	GHG tons/yr	EDX ¢/mi	Green Score	Class Ranking
STANDARD PICKUPS (cont.)									
GMC C1500 SIERRA (cont.)									
5.7L 8, auto	LEV	14	18	$1,080	$250	17	3.42	**14**	○
4.8L 8, manual	Tier 1	16	20	$1,010	$270	16	3.42	**14**	○
5.0L 8, auto	Tier 1	15	19	$1,080	$270	17	3.53	**13**	○
5.7L 8, auto	Tier 1	14	18	$1,080	$280	17	3.63	**12**	▽
GMC C2500 SIERRA									
5.3L 8, auto	Tier 1	15	19	$1,010	$270	16	3.49	**13**	○
6.0L 8, auto	Tier 1	12	16	$1,330	$290	20	3.97	**10**	✖
6.0L 8, auto	HDT	12	16	$1,310	$370	20	4.53	**8**	✖
GMC K1500 SIERRA									
4.3L 6, auto	LEV	16	20	$960	$230	15	3.13	**16**	▲
4.8L 8, manual	LEV	15	19	$1,080	$240	16	3.26	**15**	▲
4.8L 8, auto	LEV	15	18	$1,080	$240	16	3.28	**15**	▲
5.3L 8, auto	LEV	15	18	$1,080	$240	17	3.33	**14**	○
4.3L 6, auto	Tier 1	16	20	$960	$260	15	3.34	**14**	○
4.3L 6, manual	LEV	15	18	$1,080	$240	17	3.35	**14**	○
5.0L 8, auto	LEV	14	18	$1,080	$250	17	3.44	**13**	○
4.8L 8, manual	Tier 1	15	19	$1,080	$270	16	3.48	**13**	○
4.8L 8, auto	Tier 1	15	18	$1,080	$270	16	3.50	**13**	○
5.3L 8, auto	Tier 1	15	18	$1,080	$270	17	3.54	**13**	○
4.3L 6, manual	Tier 1	15	18	$1,080	$270	17	3.56	**13**	○
5.7L 8, auto	LEV	13	17	$1,150	$260	18	3.59	**12**	○
5.0L 8, auto	Tier 1	14	18	$1,080	$280	17	3.65	**12**	▽
5.7L 8, auto	Tier 1	13	17	$1,150	$290	18	3.80	**11**	✖
GMC K2500 SIERRA									
5.7L 8, auto	HDT	14	17	$1,190	$360	19	4.39	**8**	✖
MAZDA B4000									
4.0L 6, manual	Tier 1	17	22	$910	$200	14	2.79	**19**	▲
4.0L 6, auto	Tier 1	16	22	$960	$210	14	2.86	**18**	▲
4.0L 6, manual	Tier 1	17	22	$910	$250	14	3.13	**16**	▲
4.0L 6, auto	Tier 1	16	22	$960	$250	15	3.21	**15**	▲
4.0L 6, manual	Tier 1	17	20	$960	$250	15	3.24	**15**	▲
4.0L 6, auto	Tier 1	16	20	$960	$260	15	3.30	**14**	○

See pages 32-33 for explanation of data and symbols.

	Emission Standard	Fuel Economy City	Fuel Economy Hwy	Fuel Cost/yr	Health Cost/yr	GHG tons/yr	EDX ¢/mi	Green Score	Class Ranking
SMALL UTILITIES									
CHEVROLET TRACKER CONVERTIBLE									
1.6L 4, manual	Tier 1	25	28	$660	$170	10	2.23	**25**	▲
1.6L 4, manual 4wd	Tier 1	25	27	$660	$180	11	2.28	**25**	▲
2.0L 4, auto	Tier 1	23	25	$720	$180	11	2.37	**23**	○
2.0L 4, manual	Tier 1	23	25	$720	$180	11	2.38	**23**	○
2.0L 4, auto 4wd	Tier 1	23	25	$720	$180	11	2.40	**23**	○
2.0L 4, manual 4wd	Tier 1	22	25	$750	$180	11	2.41	**23**	○
CHEVROLET TRACKER HARDTOP									
2.0L 4, auto	Tier 1	24	26	$690	$180	11	2.35	**24**	○
2.0L 4, manual	Tier 1	23	25	$720	$180	11	2.39	**23**	○
2.0L 4, manual 4wd	Tier 1	22	25	$750	$180	12	2.42	**23**	○
2.0L 4, auto 4wd	Tier 1	23	25	$720	$190	12	2.46	**22**	○
ISUZU AMIGO									
2.2L 4, manual	LEV	21	24	$790	$170	12	2.40	**23**	○
2.2L 4, manual	TLEV	21	24	$790	$190	12	2.53	**22**	○
2.2L 4, manual	Tier 1	21	24	$790	$190	12	2.56	**21**	○
2.2L 4, manual 4wd	LEV	20	23	$820	$210	13	2.76	**19**	▽
2.2L 4, manual 4wd	TLEV	20	23	$820	$240	13	2.94	**17**	▽
3.2L 6, manual 4wd	LEV	18	21	$910	$220	14	2.95	**17**	▽
2.2L 4, manual 4wd	Tier 1	20	23	$820	$240	13	2.97	**17**	▽
3.2L 6, auto	LEV	17	21	$960	$220	15	3.01	**17**	▽
3.2L 6, auto 4wd	LEV	17	21	$960	$220	15	3.01	**17**	▽
3.2L 6, manual 4wd	TLEV	18	21	$910	$250	14	3.13	**16**	✖
3.2L 6, manual 4wd	Tier 1	18	21	$910	$250	14	3.16	**16**	✖
3.2L 6, auto	TLEV	17	21	$960	$250	15	3.19	**15**	✖
3.2L 6, auto 4wd	TLEV	17	21	$960	$250	15	3.19	**15**	✖
3.2L 6, auto	Tier 1	17	21	$960	$250	15	3.22	**15**	✖
3.2L 6, auto 4wd	Tier 1	17	21	$960	$250	15	3.22	**15**	✖
JEEP WRANGLER									
2.5L 4, manual	TLEV	18	20	$910	$200	14	2.81	**19**	▽
2.5L 4, manual	Tier 1	18	20	$910	$210	14	2.83	**18**	▽
2.5L 4, auto	TLEV	16	18	$1,010	$210	15	3.00	**17**	▽
2.5L 4, auto	Tier 1	16	18	$1,010	$220	15	3.03	**17**	▽
4.0L 6, manual	TLEV	16	19	$1,010	$250	16	3.30	**14**	✖
4.0L 6, manual	Tier 1	16	19	$1,010	$260	16	3.33	**14**	✖
4.0L 6, auto	TLEV	15	18	$1,080	$260	16	3.43	**14**	✖
4.0L 6, auto	Tier 1	15	18	$1,080	$260	16	3.46	**13**	✖

See pages 32-33 for explanation of data and symbols.

	Emission Standard	Fuel Economy City	Hwy	Fuel Cost/yr	Health Cost/yr	GHG tons/yr	EDX ¢/mi	Green Score	Class Ranking
SMALL UTILITIES (cont.)									
KIA SPORTAGE									
2.0L 4, manual	TLEV	19	23	$820	$190	13	2.63	**20**	○
2.0L 4, manual 4wd	TLEV	19	23	$820	$190	13	2.63	**20**	○
2.0L 4, manual	Tier 1	19	23	$820	$200	13	2.65	**20**	○
2.0L 4, manual 4wd	Tier 1	19	23	$820	$200	13	2.65	**20**	○
2.0L 4, auto	TLEV	19	23	$820	$200	13	2.69	**20**	○
2.0L 4, auto 4wd	TLEV	19	23	$820	$200	13	2.69	**20**	○
2.0L 4, auto	Tier 1	19	23	$820	$200	13	2.72	**19**	▽
2.0L 4, auto 4wd	Tier 1	19	23	$820	$200	13	2.72	**19**	▽
MERCEDES ML430									
4.3L 8, auto [P]	LEV	15	18	$1,220	$240	17	3.32	**14**	✖
4.3L 8, auto [P]	Tier 1	15	18	$1,220	$270	17	3.53	**13**	✖
SUBARU FORESTER									
2.5L 4, manual	TLEV	21	27	$720	$190	12	2.43	**23**	○
2.5L 4, manual	Tier 1	21	27	$720	$190	12	2.46	**22**	○
2.5L 4, auto	TLEV	21	26	$750	$190	12	2.49	**22**	○
2.5L 4, auto	Tier 1	21	26	$750	$190	12	2.52	**22**	○
SUZUKI GRAND VITARA 4-DOOR									
2.5L 6, manual	LEV	19	22	$860	$180	13	2.52	**22**	○
2.5L 6, manual 4wd	LEV	19	21	$860	$180	13	2.56	**21**	○
2.5L 6, auto	LEV	19	21	$860	$180	13	2.58	**21**	○
2.5L 6, auto 4wd	LEV	18	20	$910	$180	14	2.61	**21**	○
2.5L 6, manual	Tier 1	19	22	$860	$200	13	2.68	**20**	○
2.5L 6, manual 4wd	Tier 1	19	21	$860	$200	13	2.72	**19**	▽
2.5L 6, auto	Tier 1	19	21	$860	$200	13	2.74	**19**	▽
2.5L 6, auto 4wd	Tier 1	18	20	$910	$200	14	2.77	**19**	▽
SUZUKI VITARA 2-DOOR									
1.6L 4, manual	Tier 1	25	28	$660	$170	10	2.23	**25**	▲
1.6L 4, manual 4wd	Tier 1	25	27	$660	$180	11	2.28	**25**	▲
1.6L 4, auto	Tier 1	25	27	$660	$180	11	2.29	**25**	▲
1.6L 4, auto 4wd	Tier 1	24	27	$690	$180	11	2.32	**24**	○
2.0L 4, auto	Tier 1	23	25	$720	$180	11	2.40	**23**	○
2.0L 4, auto 4wd	Tier 1	23	25	$720	$180	11	2.40	**23**	○
2.0L 4, manual	Tier 1	22	24	$750	$190	12	2.44	**23**	○
2.0L 4, manual 4wd	Tier 1	22	24	$750	$190	12	2.44	**23**	○

See pages 32-33 for explanation of data and symbols.

	Emission Standard	Fuel Economy City	Fuel Economy Hwy	Fuel Cost/yr	Health Cost/yr	GHG tons/yr	EDX ¢/mi	Green Score	Class Ranking
SMALL UTILITIES (cont.)									
SUZUKI VITARA 4-DOOR									
2.0L 4, auto	Tier 1	23	25	$720	$180	11	2.41	**23**	○
2.0L 4, manual	Tier 1	22	25	$750	$190	12	2.43	**23**	○
2.0L 4, manual 4wd	Tier 1	22	25	$750	$190	12	2.43	**23**	○
2.0L 4, auto 4wd	Tier 1	23	25	$720	$190	12	2.47	**22**	○
TOYOTA RAV4									
Electric [26]	ZEV	3.2	2.6	$520	$70	5	1.05	**50**	✔
2.0L 4, manual	Tier 1	24	29	$660	$170	10	2.24	**25**	▲
2.0L 4, auto	Tier 1	24	29	$660	$180	11	2.28	**25**	▲
2.0L 4, manual 4wd	Tier 1	22	25	$750	$190	12	2.44	**23**	○
2.0L 4, auto 4wd	Tier 1	22	26	$720	$190	12	2.46	**22**	○
TOYOTA RAV4 SOFT TOP									
2.0L 4, manual	Tier 1	24	29	$660	$170	10	2.24	**25**	▲
2.0L 4, auto	Tier 1	24	29	$660	$180	11	2.28	**25**	▲
2.0L 4, auto 4wd	Tier 1	22	26	$720	$180	11	2.38	**23**	○
2.0L 4, manual 4wd	Tier 1	22	26	$720	$180	11	2.39	**23**	○
MEDIUM UTILITIES									
ACURA SLX									
3.5L 6, auto	TLEV	15	19	$1,010	$260	16	3.43	**14**	✖
3.5L 6, auto	Tier 1	15	19	$1,010	$270	16	3.46	**13**	✖
CHEVROLET BLAZER									
4.3L 6, manual	Tier 1	17	23	$910	$250	14	3.13	**16**	○
4.3L 6, auto	Tier 1	16	21	$960	$250	15	3.25	**15**	○
4.3L 6, auto 4wd	Tier 1	16	20	$960	$260	15	3.34	**14**	▽
4.3L 6, manual 4wd	Tier 1	15	18	$1,080	$270	17	3.56	**13**	✖
FORD EXPLORER									
4.0L 6, manual	LEV	18	23	$860	$220	14	2.89	**18**	▲
4.0L 6, auto 5 sp	LEV	16	21	$960	$230	15	3.07	**16**	○
4.0L 6, manual 4wd	LEV	16	20	$960	$230	15	3.07	**16**	○
4.0L 6, auto 5 sp	LEV	16	20	$960	$230	15	3.09	**16**	○
4.0L 6, auto 4wd 5 sp	LEV	15	19	$1,010	$230	16	3.19	**15**	○
4.9L 8, auto 4 sp	LEV	14	19	$1,080	$240	17	3.36	**14**	▽
4.9L 8, auto 4wd 4 sp	LEV	14	19	$1,080	$240	17	3.39	**14**	▽

[26] *Electric vehicle fuel economy is given in mi/kWh.*

	Emission Standard	Fuel Economy City	Fuel Economy Hwy	Fuel Cost/yr	Health Cost/yr	GHG tons/yr	EDX ¢/mi	Green Score	Class Ranking
MEDIUM UTILITIES (cont.)									
GMC JIMMY									
4.3L 6, manual	Tier 1	17	23	$910	$250	14	3.13	**16**	○
4.3L 6, auto	Tier 1	16	21	$960	$250	15	3.25	**15**	○
4.3L 6, auto 4wd	Tier 1	16	20	$960	$260	15	3.34	**14**	▽
4.3L 6, manual 4wd	Tier 1	15	18	$1,080	$270	17	3.56	**13**	✖
HONDA PASSPORT									
3.2L 6, manual	LEV	18	21	$910	$220	14	2.95	**17**	○
3.2L 6, manual 4wd	LEV	18	20	$910	$220	14	2.95	**17**	○
3.2L 6, auto	LEV	16	20	$960	$220	15	3.04	**17**	○
3.2L 6, auto 4wd	LEV	16	20	$960	$220	15	3.04	**17**	○
3.2L 6, manual	Tier 1	18	21	$910	$250	14	3.16	**16**	○
3.2L 6, manual 4wd	Tier 1	18	20	$910	$250	14	3.16	**16**	○
3.2L 6, auto	Tier 1	16	20	$960	$250	15	3.25	**15**	○
3.2L 6, auto 4wd	Tier 1	16	20	$960	$250	15	3.25	**15**	○
INFINITI QX4									
3.3L 6, auto	LEV	15	19	$1,080	$240	16	3.26	**15**	▽
3.3L 6, auto	Tier 1	15	19	$1,080	$270	16	3.48	**13**	✖
ISUZU RODEO									
2.2L 4, manual	LEV	21	24	$790	$170	12	2.40	**23**	✔
2.2L 4, manual	TLEV	21	24	$790	$190	12	2.53	**22**	✔
2.2L 4, manual	Tier 1	21	24	$790	$190	12	2.56	**21**	✔
3.2L 6, manual	LEV	18	21	$910	$220	14	2.95	**17**	○
3.2L 6, manual 4wd	LEV	18	20	$910	$220	14	2.95	**17**	○
3.2L 6, auto	LEV	16	20	$960	$220	15	3.04	**17**	○
3.2L 6, auto 4wd	LEV	16	20	$960	$220	15	3.04	**17**	○
3.2L 6, manual	TLEV	18	21	$910	$250	14	3.13	**16**	○
3.2L 6, manual 4wd	TLEV	18	20	$910	$250	14	3.13	**16**	○
3.2L 6, manual	Tier 1	18	21	$910	$250	14	3.16	**16**	○
3.2L 6, manual 4wd	Tier 1	18	20	$910	$250	14	3.16	**16**	○
3.2L 6, auto	TLEV	16	20	$960	$250	15	3.22	**15**	○
3.2L 6, auto 4wd	TLEV	16	20	$960	$250	15	3.22	**15**	○
3.2L 6, auto	Tier 1	16	20	$960	$250	15	3.25	**15**	○
3.2L 6, auto 4wd	Tier 1	16	20	$960	$250	15	3.25	**15**	○
ISUZU TROOPER									
3.5L 6, manual	TLEV	16	19	$1,010	$260	16	3.35	**14**	▽
3.5L 6, manual	Tier 1	16	19	$1,010	$260	16	3.38	**14**	▽

See pages 32-33 for explanation of data and symbols.

	Emission Standard	Fuel Economy City	Fuel Economy Hwy	Fuel Cost/yr	Health Cost/yr	GHG tons/yr	EDX ¢/mi	Green Score	Class Ranking
MEDIUM UTILITIES (cont.)									
ISUZU TROOPER (cont.)									
3.5L 6, auto	TLEV	15	19	$1,010	$260	16	3.43	**14**	✖
3.5L 6, auto	Tier 1	15	19	$1,010	$270	16	3.46	**13**	✖
JEEP CHEROKEE									
2.5L 4, manual	TLEV	21	25	$750	$190	12	2.51	**22**	✔
2.5L 4, manual	Tier 1	21	25	$750	$190	12	2.53	**22**	✔
4.0L 6, manual	TLEV	18	24	$860	$200	13	2.68	**20**	▲
4.0L 6, manual	Tier 1	18	24	$860	$200	13	2.71	**20**	▲
2.5L 4, auto 3 sp	TLEV	18	22	$910	$200	14	2.76	**19**	▲
4.0L 6, manual 4wd	TLEV	17	22	$910	$200	14	2.78	**19**	▲
2.5L 4, auto 3 sp	Tier 1	18	22	$910	$200	14	2.79	**19**	▲
2.5L 4, manual 4wd	TLEV	18	20	$910	$200	14	2.81	**19**	▲
4.0L 6, manual 4wd	Tier 1	17	22	$910	$210	14	2.81	**19**	▲
2.5L 4, manual 4wd	Tier 1	18	20	$910	$210	14	2.83	**18**	▲
4.0L 6, auto 4 sp	TLEV	16	22	$960	$200	14	2.86	**18**	▲
4.0L 6, auto 4 sp	Tier 1	16	22	$960	$210	14	2.88	**18**	▲
4.0L 6, auto 4wd 4 sp	TLEV	16	21	$960	$210	15	2.91	**18**	○
4.0L 6, auto 4wd 4 sp	Tier 1	16	21	$960	$210	15	2.94	**17**	○
4.0L 6, auto 4wd 4 sp	TLEV	16	21	$960	$250	15	3.25	**15**	▽
4.0L 6, auto 4wd 4 sp	Tier 1	16	21	$960	$260	15	3.28	**15**	▽
JEEP GRAND CHEROKEE									
4.0L 6, auto	Tier 1	16	21	$960	$260	15	3.27	**15**	▽
4.0L 6, auto 4wd	Tier 1	16	21	$960	$260	15	3.28	**15**	▽
4.7L 8, auto 4wd	Tier 1	15	19	$1,080	$270	17	3.52	**13**	✖
LAND ROVER DISCOVERY									
3.9L 8, auto [P]	Tier 1	14	17	$1,300	$280	18	3.66	**12**	✖
3.9L 8, auto [P]	Tier 1	13	16	$1,390	$290	19	3.84	**11**	✖
MERCEDES ML320									
3.2L 6, auto [P]	LEV	17	21	$1,080	$230	15	3.12	**16**	○
3.2L 6, auto [P]	Tier 1	17	21	$1,080	$260	15	3.33	**14**	▽
MERCURY MOUNTAINEER									
4.0L 6, auto	LEV	15	20	$1,010	$230	16	3.19	**15**	○
4.0L 6, auto 4wd	LEV	15	19	$1,010	$240	16	3.22	**15**	○
4.9L 8, auto	LEV	14	19	$1,080	$240	17	3.34	**14**	▽
4.9L 8, auto 4wd	LEV	14	19	$1,080	$240	17	3.37	**14**	▽

See pages 32-33 for explanation of data and symbols.

	Emission Standard	Fuel Economy City	Fuel Economy Hwy	Fuel Cost/yr	Health Cost/yr	GHG tons/yr	EDX ¢/mi	Green Score	Class Ranking
MEDIUM UTILITIES (cont.)									
MITSUBISHI MONTERO									
2.3L 4, manual	LEV	22	24	$750	$170	12	2.36	**24**	✔
2.3L 4, manual	TLEV	22	24	$750	$190	12	2.49	**22**	✔
2.3L 4, manual	Tier 1	22	24	$750	$190	12	2.52	**22**	✔
3.0L 6, auto	LEV	18	21	$860	$220	14	2.89	**18**	▲
3.0L 6, auto 4wd	LEV	18	21	$910	$220	14	2.93	**17**	○
3.0L 6, manual 4wd	LEV	17	21	$910	$220	15	2.97	**17**	○
3.5L 6, auto	LEV	17	20	$960	$220	15	3.01	**17**	○
3.0L 6, auto	TLEV	18	21	$860	$240	14	3.07	**16**	○
3.0L 6, auto	Tier 1	18	21	$860	$250	14	3.10	**16**	○
3.0L 6, auto 4wd	TLEV	18	21	$910	$240	14	3.11	**16**	○
3.5L 6, auto 4wd	LEV	16	20	$960	$230	15	3.13	**16**	○
3.0L 6, auto 4wd	Tier 1	18	21	$910	$250	14	3.14	**16**	○
3.0L 6, manual 4wd	TLEV	17	21	$910	$250	15	3.15	**16**	○
3.0L 6, manual 4wd	Tier 1	17	21	$910	$250	15	3.18	**15**	○
3.5L 6, auto 4wd	LEV	16	19	$1,010	$230	16	3.18	**15**	○
3.5L 6, auto	TLEV	17	20	$960	$250	15	3.19	**15**	○
3.5L 6, auto	Tier 1	17	20	$960	$250	15	3.22	**15**	○
3.0L 6, manual 4wd	LEV	16	18	$1,010	$240	16	3.24	**15**	○
3.0L 6, auto 4wd	TLEV	16	20	$960	$260	15	3.30	**14**	▽
3.0L 6, auto 4wd	Tier 1	16	20	$960	$260	15	3.33	**14**	▽
3.5L 6, auto 4wd	TLEV	16	19	$1,010	$260	16	3.36	**14**	▽
3.5L 6, auto 4wd	Tier 1	16	19	$1,010	$260	16	3.40	**14**	▽
3.0L 6, manual 4wd	TLEV	16	18	$1,010	$260	16	3.42	**14**	✖
3.0L 6, manual 4wd	Tier 1	16	18	$1,010	$270	16	3.45	**13**	✖
MITSUBISHI NATIVA									
2.3L 4, manual	Tier 1	22	24	$750	$190	12	2.52	**22**	✔
2.3L 4, manual	Tier 1	22	24	$750	$240	12	2.86	**18**	▲
2.3L 4, auto	Tier 1	19	24	$820	$240	13	3.01	**17**	○
3.0L 6, auto	Tier 1	18	21	$860	$250	14	3.10	**16**	○
3.0L 6, auto 4wd	Tier 1	18	21	$910	$250	14	3.13	**16**	○
3.0L 6, manual 4wd	Tier 1	17	21	$910	$250	15	3.18	**15**	○
3.0L 6, auto 4wd	Tier 1	18	21	$910	$250	14	3.19	**15**	○
NISSAN PATHFINDER									
3.3L 6, manual	LEV	17	19	$960	$220	15	3.06	**16**	○
3.3L 6, auto	LEV	16	20	$1,010	$230	15	3.11	**16**	○
3.3L 6, manual 4wd	LEV	16	18	$1,010	$230	16	3.22	**15**	○
3.3L 6, auto 4wd	LEV	15	19	$1,080	$240	16	3.26	**15**	▽

See pages 32-33 for explanation of data and symbols.

	Emission Standard	Fuel Economy City	Fuel Economy Hwy	Fuel Cost/yr	Health Cost/yr	GHG tons/yr	EDX ¢/mi	Green Score	Class Ranking
MEDIUM UTILITIES (cont.)									
NISSAN PATHFINDER (cont.)									
3.3L 6, manual	Tier 1	17	19	$960	$260	15	3.27	**15**	▽
3.3L 6, auto	Tier 1	16	20	$1,010	$260	15	3.32	**14**	▽
3.3L 6, manual 4wd	Tier 1	16	18	$1,010	$270	16	3.43	**13**	✖
3.3L 6, auto 4wd	Tier 1	15	19	$1,080	$270	16	3.48	**13**	✖
OLDSMOBILE BRAVADA									
4.3L 6, auto	Tier 1	16	20	$960	$260	15	3.34	**14**	▽
TOYOTA 4RUNNER									
2.7L 4, manual	Tier 1	18	23	$860	$200	13	2.69	**20**	▲
2.7L 4, auto	Tier 1	20	24	$790	$240	13	2.94	**17**	○
2.7L 4, manual	Tier 1	18	23	$860	$240	14	3.04	**16**	○
2.7L 4, auto 4wd	Tier 1	18	21	$910	$250	14	3.13	**16**	○
3.4L 6, auto	Tier 1	17	21	$910	$250	14	3.17	**15**	○
2.7L 4, manual 4wd	Tier 1	17	21	$910	$250	15	3.18	**15**	○
3.4L 6, auto 4wd	Tier 1	17	20	$960	$250	15	3.25	**15**	○
3.4L 6, manual 4wd	Tier 1	17	19	$960	$250	15	3.26	**15**	▽
3.4L 6, auto 4wd	Tier 1	17	20	$960	$260	15	3.30	**14**	▽
LARGE UTILITIES									
CADILLAC ESCALADE									
5.7L 8, auto	LEV	12	16	$1,230	$310	20	4.09	**10**	○
5.7L 8, auto	Tier 1	12	16	$1,230	$360	20	4.43	**8**	▽
CHEVROLET C1500 SUBURBAN									
5.7L 8, auto	LEV	14	18	$1,150	$250	18	3.53	**13**	▲
5.7L 8, auto	Tier 1	14	18	$1,150	$280	18	3.74	**12**	○
CHEVROLET C1500 TAHOE									
5.7L 8, auto	LEV	14	18	$1,150	$250	18	3.49	**13**	▲
5.7L 8, auto	Tier 1	14	18	$1,150	$280	18	3.70	**12**	▲
CHEVROLET C2500 SUBURBAN									
5.7L 8, auto	Tier 1	14	18	$1,150	$280	18	3.74	**12**	○
CHEVROLET K1500 TAHOE									
5.7L 8, auto	LEV	12	16	$1,230	$260	19	3.76	**12**	○
5.7L 8, auto	Tier 1	12	16	$1,230	$300	19	3.97	**10**	○
5.7L 8, auto	LEV	12	16	$1,230	$310	20	4.09	**10**	○
5.7L 8, auto	Tier 1	12	16	$1,230	$360	20	4.43	**8**	▽
6.5L 8, auto [D]	Tier 1-D	15	19	$1,030	$430	16	4.48	**8**	✖

See pages 32-33 for explanation of data and symbols.

	Emission Standard	Fuel Economy City	Hwy	Fuel Cost/yr	Health Cost/yr	GHG tons/yr	EDX ¢/mi	Green Score	Class Ranking
LARGE UTILITIES (cont.)									
CHEVROLET K2500 SUBURBAN									
5.7L 8, auto	Tier 1	14	18	$1,150	$280	18	3.74	**12**	○
FORD EXPEDITION									
4.6L 8, auto	LEV	13	18	$1,150	$250	18	3.50	**13**	▲
5.4L 8, auto	LEV	13	18	$1,230	$260	19	3.66	**12**	▲
4.6L 8, auto 4wd	LEV	12	16	$1,230	$310	20	4.09	**10**	○
5.4L 8, auto 4wd	LEV	12	16	$1,330	$310	20	4.20	**9**	○
GMC C1500 SUBURBAN									
5.7L 8, auto	LEV	14	18	$1,150	$250	18	3.53	**13**	▲
5.7L 8, auto	Tier 1	14	18	$1,150	$280	18	3.74	**12**	○
GMC C1500 YUKON									
5.7L 8, auto	LEV	14	18	$1,150	$250	18	3.49	**13**	▲
5.7L 8, auto	Tier 1	14	18	$1,150	$280	18	3.70	**12**	▲
GMC C2500 SUBURBAN									
5.7L 8, auto	HDT	14	18	$1,150	$360	18	4.26	**9**	▽
GMC K2500 SUBURBAN									
5.7L 8, auto	HDT	14	18	$1,150	$360	18	4.26	**9**	▽
GMC K1500 YUKON									
5.7L 8, auto	LEV	12	16	$1,230	$260	19	3.76	**12**	○
5.7L 8, auto	Tier 1	12	16	$1,230	$300	19	3.97	**10**	○
5.7L 8, auto	Tier 1	12	16	$1,230	$360	20	4.43	**8**	▽
6.5L 8, auto [D]	Tier 1-D	15	19	$1,030	$430	16	4.48	**8**	✖
LAND ROVER RANGE ROVER									
3.9L 8, auto [P]	Tier 1	13	16	$1,390	$290	19	3.84	**11**	○
4.6L 8, auto [P]	Tier 1	13	16	$1,390	$290	19	3.88	**11**	○
LEXUS LX 470									
4.7L 8, auto	Tier 1	13	16	$1,150	$350	19	4.31	**9**	▽
LINCOLN NAVIGATOR									
5.4L 8, auto [P]	LEV	13	17	$1,300	$260	18	3.63	**12**	▲
5.4L 8, auto	LEV	13	18	$1,230	$260	19	3.66	**12**	▲
5.4L 8, auto [P] 4wd	LEV	12	16	$1,390	$310	20	4.11	**10**	○
5.4L 8, auto 4wd	LEV	12	16	$1,330	$310	20	4.20	**9**	○
TOYOTA LAND CRUISER WAGON									
4.7L 8, auto	Tier 1	13	16	$1,150	$350	19	4.31	**9**	▽

See pages 32-33 for explanation of data and symbols.

Driving Green

Buying Green is just the first step in reducing the environmental impacts of automobile use. Your choice of vehicle is most important, but how you drive and how well you maintain your car, van, or light truck will also make a difference.

Drive Carefully and Gently

- Avoid "jack rabbit" starts and aggressive driving. Flooring the gas pedal not only wastes gas, it leads to drastically higher pollution rates. One second of high-powered driving can produce nearly the same volume of carbon monoxide emissions as a half hour of normal driving.
- Think ahead. Try to anticipate stops and let your vehicle coast down as much as possible. Avoid the increased pollution, wasted gas, and wear on your brakes created by accelerating hard and braking hard.
- Follow the speed limit! Driving 75 mph instead of 65 mph will lower your fuel economy by about 10 percent, and can dramatically increase tailpipe pollution in many vehicles.
- When possible, plan your trips to avoid rush hour. Stop-and-go driving burns gas and increases emissions of smog-forming pollutants.
- Combine trips. Warmed-up engines and catalysts generate much less air pollution, so combining several short trips into one can make a big difference.
- Take a load off. Carrying around an extra 100 pounds reduces fuel economy by about 1 percent. Take a few moments to unload your cargo area.
- If your vehicle has it, use overdrive gear at cruising speeds. When driving a manual transmission, shift up as soon as possible. Running in a higher gear decreases the rpm and will decrease fuel use and engine wear.
- Try using the vents and opening windows to cool off before you turn on the air conditioner. Air conditioner use increases fuel

Whatever Vehicle You Use, It's Greener with Two Passengers in It Than It Is with One, and It's Greener Still with Three.

The *Green Guide to Cars and Trucks* ratings are based on a given vehicle's environmental performance. But it's also fair to look at greenness in terms of how a vehicle is used. Pollution per person is lower when more than one passenger is carried. For example, if you regularly carry three in a car or van, you can think of your household's per-person pollution as being one-third lower than that of someone who mostly drives alone in the same model.

consumption, increases NO_x emissions in some vehicles, and involves environmentally damaging fluids.

- Unlike many older cars and trucks, modern vehicles don't need to warm up and they have automatic chokes, so you don't need to step on the gas pedal before starting the engine.

Maintenance Tips

Your owner's manual is the best source of information about how to care for your vehicle. Refer to it for instructions about proper maintenance and service specific to your make and model.

- Keep your tires properly inflated. Tires should be inflated to the pressure recommended for your vehicle; this information is often printed inside the door frame or in your owner's manual. For every 3 pounds below recommended pressure, fuel economy goes down by about 1 percent.
- Check your own fuel economy every few weeks. If you notice it slipping, that could mean you have a minor problem with the engine or your brakes. Using this advance warning, you can fix problems before you have a breakdown on the road.
- Get a tune-up. Whether you do it yourself or go to a mechanic, a tune-up can increase your fuel economy. Follow owner's manual guidelines. Be sure to check for worn spark plugs, dragging brakes, and low transmission fluid, have your wheels aligned and tires rotated, and replace air filter if needed. Make sure all used vehicle fluids are recycled or disposed of safely.
- Change the oil. In addition to making your car or truck last longer,

Checking Your Gas Mileage

You can check your MPG for yourself with these three easy steps:

1. Fill the tank and jot down the odometer reading.
2. Next time you buy gas, fill the tank again and note the new odometer reading and the number of gallons you purchased.
3. Subtract the old odometer reading from the new one and divide this number by the number of gallons you bought.

The result is your miles per gallon (MPG).

replacing the oil and oil filter regularly will also help fuel economy. Check your owner's manual for specific recommendations about how often to change. Ask the service station if they recycle used oil, or if you do it yourself, take your old oil to someplace that does recycle. Ask for recycled oil as a replacement.

- Have your vehicle's emission control system checked periodically. Take it in for service if an instrument panel warning light comes on.

Careful Fill-Ups

Americans too often take gasoline for granted, forgetting that it is quite a hazardous substance. Gasoline fumes are toxic and carcinogenic; they cause smog; and spilled gasoline can pollute the water and poison wildlife. And it's very flammable, too.

- Use regular gasoline unless your owner's manual says otherwise. Unless your car requires premium, high-octane fuels improve neither fuel economy nor performance and will just waste your money.
- Don't overfill the gas tank or try to top it off beyond where the automatic nozzle clicks off. Spilled gasoline evaporates to aggravate smog formation and can leak into groundwater.
- Patronize gas stations that have vapor-recovery nozzles (those black, accordion-looking plastic devices attached to the nozzle) whenever you can.

Prudent Parking

- Park in the shade in summer to keep your car cool and minimize evaporation of fuel.

- If you have a garage, use it as much as possible to keep your car warm in winter and cool in summer.
- If you have to park outdoors, windshield shades can cut down on summer heat and help keep the frost off in the winter.

Use "Commuter Choice" Programs

Do you commute to work? If so, you might be eligible for cash benefits for greener commuting. Through Commuter Choice programs, employers can offer federal tax free transit or vanpool benefits to their employees up to certain limits (e.g., $65/month). Employers can also allow employees to choose to receive taxable cash instead of parking benefits (up to $175/month). Employees can use this cash to commute as they wish, including carpooling, telework, bicycling, or walking. Employers benefit through lowered taxes, lowered costs, and enhanced recruitment and retention. Commuter Choice programs have the potential to reduce automobile pollution, reduce greenhouse gasses, reduce traffic congestion, and conserve fossil fuels and energy. Ask your employer if they have a Commuter Choice program. If not, ask them to start one.

Information on Commuter Choice programs is available on the web at: http://www.epa.gov/omswww/traq

Automobiles and the Environment

Automobiles affect the environment in many ways. Impacts begin when a vehicle is manufactured (including the production of the all the parts and materials that go into the car) and end with its scrappage in a junkyard (which can recycle many parts but also involves the disposal of many wastes). Over the life of an average motor vehicle, however, much of the environmental damage occurs during driving and is greatly associated with fuel consumption. The figure below illustrates the breakdown of lifecycle ("cradle to grave") energy use for a typical automobile. Nearly 90 percent is due to fuel consumption over the dozen or so years of a vehicle's life.

Environmental impacts start with mineral extraction and the production of the raw materials that go into the parts of a car. For example, iron ore gets turned into steel, which now accounts for most of the mass in vehicles. Steel can be recycled, of course. On average, today's automobiles are about 75 percent recyclable, and using recycled steel helps reduce energy use and pollution. Other metal components, such as aluminum (used in some engine parts and wheels, for example) and copper (used for wiring) are also largely recycled. The lead and acid in batteries are poisonous and dangerous. But batteries can be recycled, if they are returned to a service station, a parts store, or brought to a municipal hazardous waste facility. Plastics, which are mostly made from petroleum, are more difficult to recycle. In any case, some degree of pollution is associated with all of these components, much of it due to the energy consumption, air pollution, and releases of toxic substances that occur when automobiles are manufactured and distributed.

Most of the environmental impact associated with motor vehicles occurs when they are used, due to pollution in their exhaust and pollution associated with supplying the fuel. In the United States, nearly all of today's automobiles use gasoline; a lesser number use diesel fuel. In some areas, various alternative fuels are being introduced, but these are not widely available for most drivers. When gasoline, diesel, or other fuels are burned in car engines, combustion is never perfect, and so a mix of hazardous pollutants comes out the tailpipe.

If combustion were perfect and didn't create noxious byproducts, the exhaust would contain only water vapor and carbon dioxide. Carbon dioxide (CO_2) isn't directly harmful to health, at least

LIFECYCLE ENERGY USE FOR A TYPICAL AUTOMOBILE

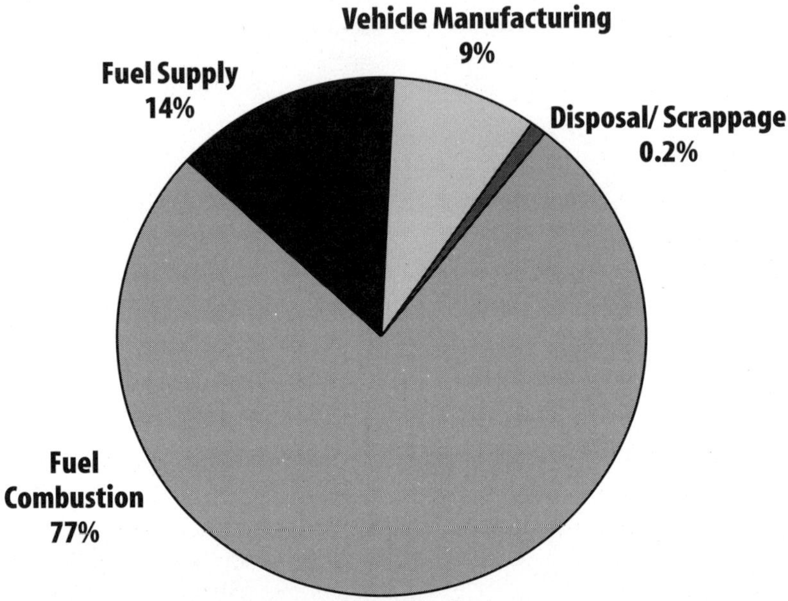

not in low concentrations. After all, CO_2 is also what we exhale as a result of our "burning" the calories in the food we eat. However, CO_2 from fossil fuels like gasoline and diesel is very harmful to the environment because it causes global warming—more on this pollutant shortly.

Motor fuel is itself a product and so, like a car, environmental damage occurs throughout its lifecycle as well. For gasoline and diesel, the product lifecycle begins at the oil well and ends when the fuel is burned in the engine. *Fuel cycle* impacts are the forms of pollution and other environmental damage that occur between the oil well and the fuel tank. Gasoline and diesel fuel are poisonous to humans, plants, and animals, and their vapors are toxic. Other energy sources have their own fuel cycles. With battery-powered electric vehicles, for example, no fuel is burned onboard the vehicle, and so nearly all of the fuel-cycle pollution and energy use occurs at electric power plants and in producing the fuels that run the power plants. Many of the same air pollutants that spew from vehicle tailpipes are also spewed from power plants and oil refineries (as well as the tanker trucks that deliver gasoline to your local filling station).

Gasoline and diesel fuel now provide 97 percent of America's transportation energy needs. Air pollution isn't the only problem associated with these petroleum-based fuels. Oil extraction lays waste to many fragile ecosystems, harming tropical forests in South America and Southeast Asia, deserts and wetlands in the Middle East, our own coastal areas, and the fragile tundra and arctic coastal plains of Alaska. Millions of gallons of oil are spilled every year. Sometimes the disasters are well known, such as the 1989 Exxon Valdez spill in Prince William Sound. More often there are rarely reported but still tragic smaller spills that occur in the oceans and in coastal waters, bays, and rivers throughout the world. In our own communities, groundwater is sometimes tainted by leaks from underground fuel storage tanks and miscellaneous spills that occur during shipping and handling of the 120 billion gallons of fuel we use each year.

In addition to these environmental harms, gasoline and diesel consumption bring economic and security risks. The Middle East contains the largest concentration of the world's oil. The United States maintains a global military presence partly to maintain access to foreign oil. Most recently in the 1991 war with Iraq, but several times throughout our history, U.S. troops were committed and blood was spilled to secure our oil supply.

Major recessions were triggered by oil crises in the 1970s and early 1980s, causing unemployment and inflation. Oil imports drain $50 billion per year from American pockets, representing lost job opportunities even when our economy seems to be doing fine. Half of U.S. oil is now imported and our dependence on foreign sources is steadily rising, perpetuating the risk of future oil crises. Moreover, our addiction to gasoline and diesel fuel involves moral compromises through association with some oil-rich countries whose standards of human rights and environmental protection are not the same as what we expect at home. Of course, these issues go beyond strictly environmental concerns. Nevertheless, choosing greener vehicles that consume less fuel will not only protect the environment, but also will help strengthen our economy while reducing the security risks and moral liabilities of oil dependence.

Major Pollutants Associated with Automobiles

Our focus in the *Green Guide to Cars and Trucks* is on air pollutants related to car and truck fuel consumption, because they are such a large part of a vehicle's environmental damage and because they are the main impacts that can be reduced through your choice of make and model. The adjoining figure shows the amounts of major air pollutants caused by the average new passenger car and light truck in a year. The pollution coming from vehicles can differ depending on the standards they meet (and how well their emissions controls work), how they are driven and maintained, and the quantity and quality of the fuel they burn. Many vans, pickups, sport utilities, and other light trucks meet less stringent emissions and fuel economy standards than vehicles classified as passenger cars. As a result, and as the ratings in this guide indicate, the average light truck pollutes more than the average car.

All new vehicles must meet either the emissions standards set by the U.S. Environmental Protection Agency (EPA) or those set by the California Air Resources Board (CARB). Generally, California standards are more stringent than the Federal standards. Massachusetts and New York have adopted the California standards and other Northeastern states are also considering them. In many states, however, only EPA-certified vehicles may be available.

Vehicles are also tested for fuel economy, as measured by miles per gallon—MPG. Fuel economy standards apply to manufacturers, rather than individual vehicles, and are set so that an automaker can sell a variety of vehicles as long as the average MPG of its sales meets the applicable standard. Manufacturers calculate the fuel economy of each model they sell using laboratory tests similar to those conducted to determine tailpipe emissions. Because these tests give fuel economy estimates higher than what most people experience in real-world driving, the MPG measurements are adjusted downward by EPA. These adjusted MPG numbers are printed on new vehicle stickers and listed here in the *Guide*.

Although a wide variety of pollutants are formed in the various stages of an automobiles lifecycle, our ratings are mainly based on the serious air pollutants that are regulated to control vehicle pollution. All of these pollutants are more damaging to health when emitted from vehicle tailpipes than when a similar quantity is emitted from a power plant, since tailpipe pollution is literally "in your face," subjecting people to more direct exposures during daily activities.

Particulate Matter (PM)

Fine airborne particles are an established cause of lung problems, from shortness of breath to worsening of respiratory and cardiovascular disease, damage to lung tissues, and cancer. Certain people are

POLLUTION FROM A TYPICAL NEW CAR AND LIGHT TRUCK

Pounds of pollutant per year, including both tailpipe and fuel-cycle emissions and assuming it is driven 15,000 miles per year.

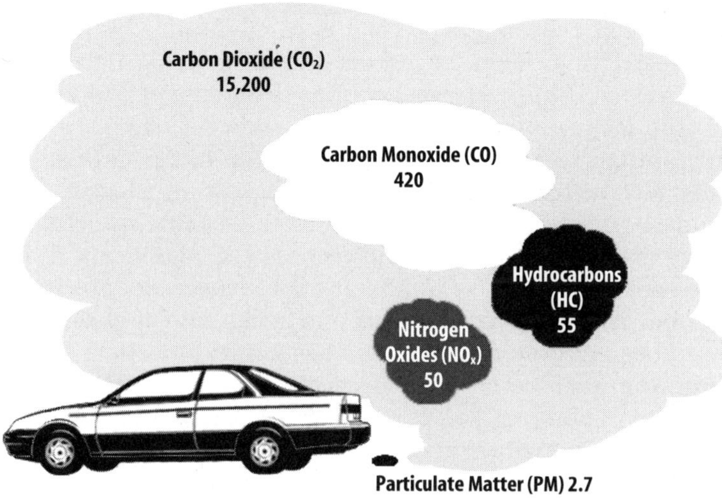

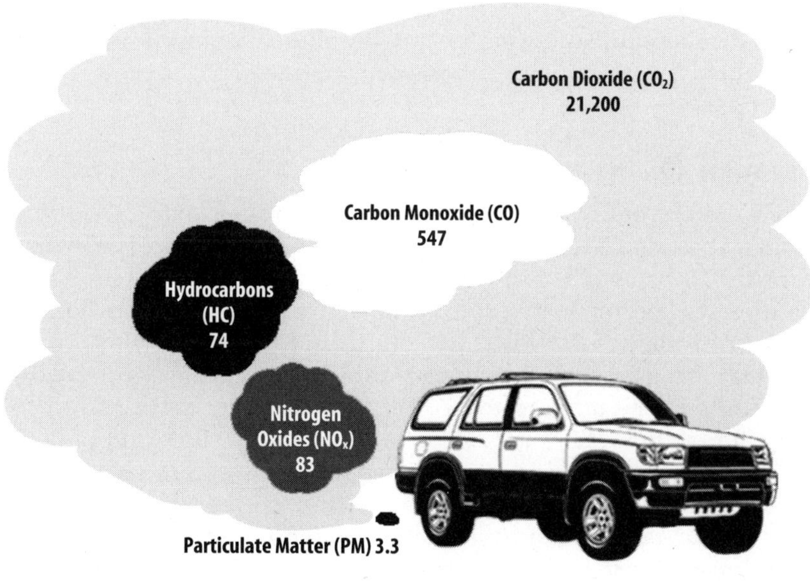

The average light truck pollutes 40 percent more than the average car.

particularly vulnerable to breathing air polluted by fine particles, among them asthmatics, individuals with the flu and with chronic heart or lung diseases, as well as children and the elderly. PM also soils and damages buildings and materials. It also forms haze that obscures visibility in many regions. Soot and smoke coming from exhaust pipes are obvious sources of PM, but among the most deadly forms of airborne particulate matter are the invisible fine particles that lodge deeply in the lungs. PM has been regulated for some time, but the regulations were based on counting all particles up to 10 microns in size (PM_{10}). However, PM_{10} standards fail to adequately control the most dangerous, very fine particles. The U.S. EPA has recently started to regulate fine particles up to 2.5 microns in size ($PM_{2.5}$), which better focuses on the most damaging category.

Properly functioning new, fuel-injected gasoline vehicles directly emit very little $PM_{2.5}$. But they indirectly cause significant PM pollution as a result of their NO_x, SO_2, and HC emissions, not only from tailpipe but also from vehicle manufacturing and fuel refining (see below). These emissions result in "secondary" particle formation. This phenomenon refers to the way that the gaseous pollutants agglomerate ("glom up") at microscopic scales to form fine particles that are largely invisible but cause the health problems mentioned. Transportation sources account for about 20 percent of directly emitted $PM_{2.5}$. Diesel engines are the major source of direct PM emissions from motor vehicles. Although most such emissions come from heavy trucks and diesel buses, even the smaller diesel engines in some cars and light trucks emit significant amounts of fine PM.

Nitrogen Oxides (NO_x)

NO_x refers mainly to two chemicals, nitrogen oxide (NO) and nitrogen dioxide (NO_2) that are formed when nitrogen gas, which comprises 78 percent of air, reacts with oxygen during the high temperatures that occur during fuel combustion. NO_x is truly a noxious pollutant in many ways. It is directly hazardous, an irritant to the lungs that can aggravate respiratory problems. It reacts with organic compounds in the air to cause ozone, which is the main reason for "smog alerts" that still happen too often in many cities and regions. NO_x is a precursor of fine particles, which cause respiratory problems and lead to thousands of premature deaths each year. It is also a precursor of acid rain, which harms lakes, waterways, forests, and other ecosystems, as well as damaging buildings and crops. Airborne NO_x also contributes to nitrification—essentially an over-fertilization—of wetlands and bays, leading to algae blooms and fish kills.

As an air pollutant, NO_x is one of the most difficult to control since it is such a pervasive product of combustion. Nationwide, most NO_x comes from electric power plants and industrial sources. Natural gas and oil-fired home furnaces and water heaters also produce NO_x in their flue gases. Motor vehicles account for about one-third of nationwide NO_x emissions. Many of these emissions come from heavy-duty diesel trucks, but cars and light trucks are also a major source. NO_x has also been one of the most difficult pollutants to get out of our air. EPA air quality regulations have helped keep emissions from growing as fast as they might have, but nationwide, overall NO_x emissions are higher than they were a decade ago. A good portion of this growth in NO_x pollution comes from cars and light trucks.

Sulfur Dioxide (SO_2)

Gasoline and diesel fuels also contain varying amounts of sulfur, which burns in the engine to produce sulfur dioxide (SO_2). This gaseous chemical is another source of secondary particulate formation, and is itself a lung irritant as well as a cause of acid rain. SO_2 also interferes with the operation of catalytic converters. Some of the cleaner, reformulated versions of gasoline have very low sulfur levels. But most gasoline sold nationwide is still quite dirty in terms of sulfur content.

Cars and light trucks are not the largest source of SO_2 emissions, which come mainly from power plants and industrial facilities. However, because cars and light trucks are so numerous and gasoline has a high average sulfur content, cars and light trucks cause twice as much fine PM pollution as heavy freight trucks. Making all gasoline as clean as the cleaner, low-sulfur fuels already available in California would greatly reduce this PM pollution from all cars and trucks on the road, both new and used.

Hydrocarbons (HC)

Hydrocarbons are a broad class of chemicals containing carbon and hydrogen. Those hydrocarbons that cause various forms of air pollution are also known as volatile organic compounds since they are forms of HC that are either gases or readily evaporate into the air. Many forms of HC are directly hazardous, contributing to what are collectively called "air toxics." These compounds can be directly irritating to the lungs and other tissues and they can also cause cancer, contribute to birth defects, and cause other illnesses. During daylight hours, and particularly during hot summer weather, HC reacts with NO_x to form ozone smog (see the box on page 82). Controlling ozone

Ozone: Helpful in the Stratosphere, But Harmful in the Air We Breathe

Ozone (O_3) is a highly reactive form of oxygen that occurs naturally in various parts of the atmosphere but gets artificially produced in dangerously high concentrations due to emissions from cars, trucks, and other combustion sources.

Up in the stratosphere, ozone helps protect us from ultraviolet radiation. Loss of this protective ozone layer at high altitudes can lead to increased skin cancer. Such concerns have led to restrictions on ozone-depleting chemicals such as those once found in some spray cans and others that have been phased out of use in refrigerators and air conditioners (including automotive air conditioners).

Down in the lower atmosphere, in the air the we breathe, ozone is a health hazard. It is the main ingredient of the smog that causes pollution alerts in many cities around the country. Ozone produced by pollution at low altitudes is of no help in restoring the protective ozone layer at high altitudes. Inhaling air polluted by ozone damages the lungs, reduces breathing ability, and makes us more susceptible to other respiratory problems. Ozone can be deadly to individuals with asthma and other lung conditions, as well as to people with heart conditions. It is also harmful to both adults and children who are otherwise healthy. The risks of shortness of breath, chest pain, lung congestion, and other symptoms caused by ozone are the reasons why public health officials warn us to stay inside and avoid strenuous exercise on severe air pollution days.

Although cars and trucks do not directly emit ozone, they are a major cause of ozone smog. They add to the amount of HC in the air, and tailpipe NO_x reacts with the HC to form ozone. Cities without major industries and power plants still have serious smog problems, mostly caused by pollution from cars, trucks, and vans. Although many U.S. cities are seeing better air quality, we'll have to do better at cutting motor vehicle pollution to ensure progress.

is one of the major environmental challenges in the United States. Although progress has been made over the past several decades, many cities and regions still have smog alerts when ozone levels get too high.

Gasoline vapor contains a mix of hydrocarbons. Thus, HC pollution is produced whenever we fill our tanks. Some regions have special nozzles on fuel pumps to help trap such vapors. Other HC vapors are released at various stages along the way from the refinery to the filling station. Vapors seep out, even when a car is parked and turned off, due to the imperfect sealing of the fuel tank, pipes and hoses, and other components leading to the engine. HC also comes out of the tailpipe, as a result of incomplete combustion and the less-than-perfect cleanup of exhaust gases by catalytic converters and other vehicle emissions controls. Diesel fuel is less volatile than gasoline, so evaporation is less of a problem. Nevertheless, diesel exhaust still contains many toxic hydrocarbons and other compounds. Overall, transportation is responsible for about 36 percent of man-made HC emissions in the United States.

Carbon Monoxide (CO)

Carbon monoxide is an odorless, colorless, but potentially deadly gas that is created by the incomplete combustion of any carbon-containing fuel, including gasoline and diesel. When inhaled, CO combines with the hemoglobin in our blood, impairing the flow of oxygen to our brain and other parts of the body. We've all heard stories of people being killed by carbon monoxide poisoning, from vehicles in closed garages, during fires, or in homes when indoor CO concentrations are raised by malfunctioning stoves or furnaces. Even if it doesn't cause death, CO exposure can cause permanent damage to the nervous system. At lower concentrations, CO is still harmful, particularly for people with heart disease. In some areas, cars and trucks can create enough CO to cause health risks outdoors.

Large amounts of CO are produced when a vehicle first starts up and its engine is cold. Poorly designed and malfunctioning engines and emission controls systems are also responsible for excess CO pollution. Motor vehicles are responsible for about 60 percent of CO emissions nationwide.

Cars, Trucks, and Global Warming

The gasoline-powered automobile was invented just over 100 years ago, when the industrial revolution was still young. Streams had long been dammed to turn mills, and coal was on its way to widespread use—it was already powering steamships and locomotives. But most energy used by humans still came from traditional fuels such as wood. In 1890, the world population was about 1.5 billion but growing rapidly. The amount of carbon dioxide (CO_2) in the atmosphere was just over 290 parts per million, not yet noticeably over its level throughout pre-industrial civilization.

The world population has now topped 5.8 billion and is still growing rapidly. During the past century, the amount of fossil fuel

The Kyoto Protocol

In December 1997 the nations of the world gathered in Kyoto, Japan, to negotiate a treaty to control emissions of greenhouse gases that are causing global warming. The resulting agreement is known as the Kyoto Protocol. In it, the United States and other wealthier, developed nations—who have produced most of the world's global warming pollution—agreed to significantly reduce their emissions over the next 10 to 15 years. Greener vehicles will be essential for accomplishing that goal.

we consume has risen nearly five times faster than population. As a result, the amount of CO_2 in the atmosphere is at 360 parts per million and climbing. This rapid increase in CO_2 concentration represents an enormous impact of our energy-consumptive lifestyle on the planet, and it is causing dangerous changes to the earth's climate. The past decade has already seen many years with above-normal temperatures. The changes in weather patterns and increases in severe events are consistent with climate disruption. Last year, 1998, was the hottest ever recorded.

Carbon dioxide is the most important of what are known as greenhouse gases, compounds that enable the earth's atmosphere to trap heat, like a greenhouse, but on a global scale. Too much greenhouse gas in the atmosphere causes global warming, an increase in global average temperatures above what they normally would be.

The risks of global warming are many. Human health is threatened by more frequent and severe heat waves and the spread of tropical diseases. Lives can be lost because of rising sea levels and more severe storms, which can also damage regional and national economies. The disruptions to climate are unpredictable but certainly risky. While some areas may see greater coastal flooding and inundating rains, other regions may experience droughts. Both agriculture and natural habitats can be harmed. Future generations will bear the brunt of these risks, but the effects of global warming have already been detected. Although we cannot attribute any given event to climate change, the increased risks have created a call for action to curtail CO_2 emissions around the world.

Oil is now the world's dominant fuel. There are 600 million cars and trucks in the world, and both here and abroad, transportation accounts for most oil use. In the United States, we now have more motor vehicles than licensed drivers, and we travel over 2 trillion miles per year, burning 120 billion gallons of gasoline. Not counting the "upstream" emissions from producing the fuel, the result is over a billion tons of CO_2 pollution each year.

U.S. cars and light trucks alone account for more energy-related CO_2 than the nationwide emissions of all but three other countries in

GLOBAL CO_2 EMISSIONS: U.S. CARS AND TRUCKS vs. NATIONS OF THE WORLD

Energy-related CO_2 emissions from countries having a 1 percent or greater share of the world total compared with CO_2 emissions from U.S. light duty vehicles (LDVs).

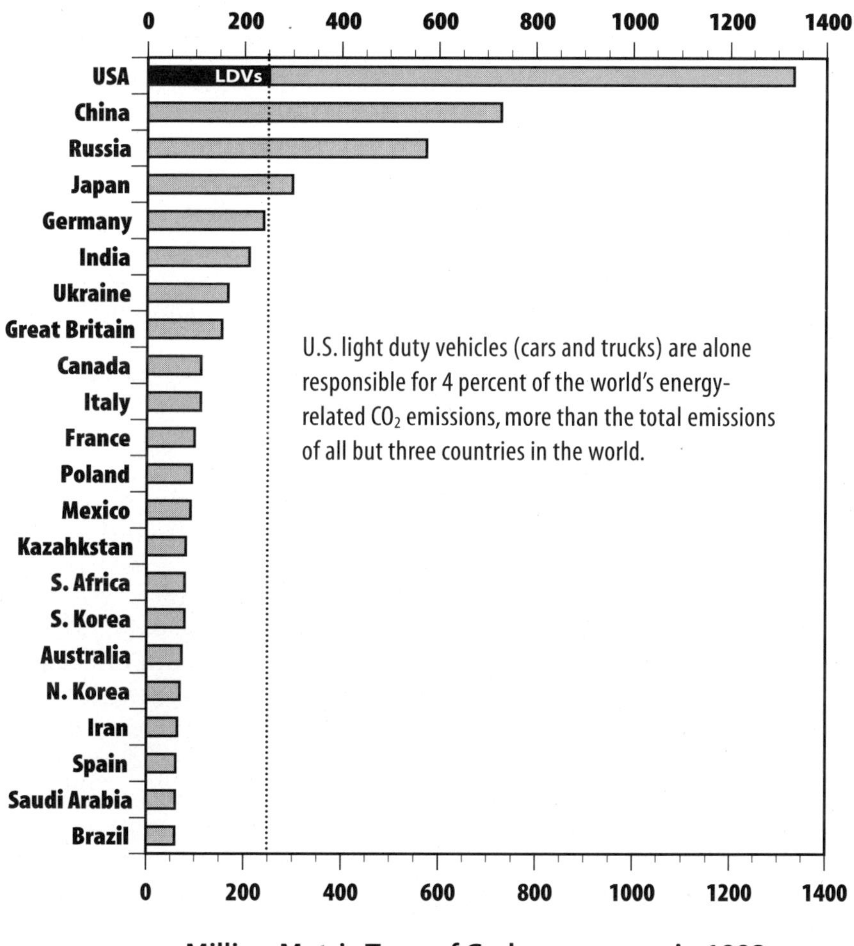

the world (China, Russia, and Japan). Our vehicles produce more CO_2 than all of India, which has more than triple our population. U.S. cars and trucks emit twice as much fossil-fuel CO_2 as the economies of either South Korea or Mexico and over three times as much as the whole of Brazil. Although some of these countries are growing and industrializing rapidly, it will be decades before their level of CO_2 pollution per person approaches ours.

Fuel Economy and Air Pollution

The amount of CO_2 emitted by a vehicle is essentially proportional to the amount of fuel burned. Thus, fuel-efficient vehicles are the best choice for helping to stop global warming. And gas guzzlers are global polluters.

For other forms of air pollution, the relation between fuel economy and emissions is more complex. Automobile emissions are regulated to a given number of grams per mile, independently of how much fuel they burn (although standards are weaker for many gas-guzzling light trucks). But several factors result in NO_x, HC, CO, and PM pollution to be higher when a vehicle's fuel economy is lower.

In real-world use, most vehicles' emissions are much higher than the standards levels. The reasons include the fact that automakers' and EPA's emissions tests fail to fully represent real-world driving, malfunction of emissions control systems, deterioration of components, inadequate or incorrect maintenance, and sometimes tampering. A portion of this excess pollution is proportional to a vehicle's rate of fuel consumption. Automobiles that meet a more stringent emissions standard are generally cleaner than those that meet a less stringent standard. However, among vehicles that meet the same standard, those with higher fuel economy generally produce less air pollution.

A significant amount of pollution also occurs in supplying vehicles with fuel. These so-called upstream emissions occur everywhere from the oil well and refinery to the filling station and gas tank, before the fuel gets to the engine. The relationship between fuel consumption and upstream emissions is strongest for hydrocarbons (HC). For an average car, about 11 grams of excess HC pollution (beyond what comes out of the tailpipe) occurs for every gallon of gasoline burned. Lesser but still-significant amounts of other pollutants are also related to the amount of fuel burned. Examples include NO_x and PM from tanker trucks delivering gasoline and a whole soup of pollutants from oil refineries. Thus, higher fuel consumption implies higher upstream pollution.

Efficiency and Safety

Other things being equal, a smaller, lighter vehicle is more fuel efficient and therefore less polluting than a larger, heavier vehicle. But are smaller, lighter vehicles less safe? The answer is more complex than one might think. In a two-car collision, occupants of the heavier vehicle are typically subjected to lower crash forces. However, the heavier vehicle will inflict higher crash forces on the occupants of the lighter vehicle. Thus, any protective benefits of mass itself, if any, come at the expense of harm to others.

While individuals may protect themselves by driving heavier vehicles, they then become a greater menace to others on the road—including pedestrians, bicyclists, and motorcyclists as well as occupants of smaller vehicles. Thus, public safety is not likely to be enhanced by a shift to heavier vehicles. Safety is enhanced by design features that improve occupant protection, such as energy-absorbing structures and passenger restraints such as seat belts. Using seat belts doubles the chance of surviving a serious crash in any vehicle. Air bags, better vehicle structures, and improved stability (designs less prone to rollover) all improve safety far more than added mass.

The fallacy of simplistic "larger is safer" thinking is well illustrated by sport utility vehicles. While large SUVs have low fatality rates (though not quite as low as some cars, in spite of the fact that they are heavier), large SUVs are deadly to other people on the road. Small SUVs tend to handle poorly, are prone to rollover, and have among the highest accident rates. Thus, SUVs are not as safe as some people believe, and neither are large pickup trucks and vans.

Evidence from accident statistics now indicates that the rising popularity of large SUVs and pickups is contributing to increased traffic fatalities and serious injuries. Because of their higher bumpers, more rigid structures, and heavier mass, these light-truck class vehicles now pose a serious threat to the occupants of passenger cars. Even though cars still outnumber light trucks on the road by two to one, many more people are killed when light trucks collide with cars than in two-car collisions. A number of insurance companies have announced higher liability premiums for some sport utilities and pickups. In short, SUVs and other large light trucks can harm both public safety and the environment, and are more costly to own and operate. Avoid these vehicles unless you truly have an ongoing need for their larger power and capacity.

Government crash-test scores measure how well a given vehicle protects its occupants. However, the scores do not tell the whole story about vehicle safety, since they fail to account for vehicle aggressivity—how harmful a vehicle is to others on the road. Nevertheless, crash-test scores are a good comparison of the relative safety of vehicles within a given size class. Well-designed vehicles that provide better crashworthiness without high aggressivity can protect their occupants without imposing greater risks on others. Crash test scores are published by the National Highway Traffic Safety Administration (NHTSA) and, along with discussions of other auto safety features, are covered by a number of publications. Advice, information on crashworthiness, and listings of safety features by make and model are provided by *The Car Book; The Truck, Van and 4×4 Book*; and *Consumer Reports.*

For Further Information

Check out the American Council for an Energy-Efficient Economy's website (http://aceee.org/greenercars) for Green Scores of new vehicles as they come on the market. We also will keep an updated Best of 1999 list available.

The following are good sources for further information about the environmental impact of vehicles and vehicle safety.

- California Air Resources Board Buyer's Guide to Cleaner Cars
  http://arbis.arb.ca.gov/msprog/ccbg/ccbg.htm
- *The Model Year 1999 Fuel Economy Guide*, published by the U.S. Department of Energy
  http://www.eren.doe.gov/feguide/
  A free printed version is available in new car showrooms or by calling 1-800-363-3732.
- The Environmental Protection Agency's Office of Mobile Sources Consumer Information Division
  http://www.epa.gov/OMSWWW/consumer.htm
- National Highway Traffic Safety Administration
  http://www.nhtsa.dot.gov/

The following public interest organizations maintain websites with information about motor vehicle environmental issues.

- The American Council for an Energy-Efficient Economy
  1001 Connecticut Avenue, N.W., Suite 801
  Washington, D.C. 20036
  http://aceee.org
- The Center for Auto Safety
  2001 "S" Street N.W., Suite 410
  Washington, D.C. 20009
  http://www.essential.org/cas/
- Campaign on Auto Pollution
  310 D Street, N.E.
  Washington, D.C. 20002
  http://www.icta.org/campaigns/cap

- Natural Resources Defense Council
  40 West 20th Street
  New York, N.Y. 10011
  http://www.nrdc.org/nrdc/camp/catran.html
- Public Citizen
  215 Pennsylvania Avenue, S.W.
  Washington, D.C. 20003
  http://www.citizen.org
- The Sierra Club
  85 Second Street, Second Floor
  San Francisco, Calif. 94105-3441
  http://www.sierraclub.org
- The Union of Concerned Scientists
  Two Brattle Square
  PO Box 9105
  Cambridge, Mass. 02238-9105
  http://www.ucsusa.org/transportation/index.html

The following printed consumer guides list fuel economy, safety and reliability information for new cars and trucks.

- *The Car Book* and *The Truck, Van and 4×4 Book* by Jack Gillis. The new editions of Jack Gillis's *Car Book* and *Truck, Van, and 4x4 Book* include Green Scores based on the tables we give here, so you can evaluate a vehicle's greeness in the context of its other attributes. These books can be ordered through the Center for Auto Safety (see p.103).
- The Consumer Reports *New Car Yearbook.* See also the Consumer's Union website: http://www.ConsumerReports.org/
- *The Autograph Book* by the American Automobile Association

Appendix: How the Ratings are Calculated

Many factors determine the environmental impact of a car or light truck. Tailpipe emissions and fuel efficiency are clearly important, but impacts also depend on the type of fuel used and the materials that go into manufacturing the vehicle. A scientific approach for estimating the environmental impacts of a product is known as *lifecycle assessment*, since it traces the impacts of a product from "cradle to grave": materials production and product manufacturing; emissions and other effects when the product is in use; through end-of-life effects of disposal and recycling. We developed the Green Scores and Class Rankings according to the principles of lifecycle assessment, using available data that are sufficiently standardized to be applicable to all makes and models.

Three types of vehicle-specific data form the basis of the *Green Guide to Cars and Trucks* ratings: tailpipe emissions, given by the emissions standard to which a vehicle is certified; fuel economy, based on EPA test cycles; and vehicle mass (curb weight).

In real-world driving, tailpipe pollution (CO, HC, NO_x, and PM) is 2 to 3 times higher than the nominal grams-per-mile (g/mi) emission standard to which a vehicle is certified. These excess emissions occur for a variety of reasons: inaccuracy of the tests, malfunctioning emission control systems; and deterioration of the catalytic converter and other components. Therefore, we apply adjustment factors, similar to those used in EPA's vehicle emissions calculation models, to determine the expected lifetime average emissions for vehicles meeting a given standard. The same adjustment factors are used for all makes and models certifying to a given emissions standard for a given fuel.

Fuel economy data are used to calculate greenhouse gas emissions, fuel-cycle criteria emissions (air pollution due to producing and distributing the fuel) and those aspects of vehicle emissions that are related to fuel consumption rates (such as a portion of evaporative HC emissions). Fuel economy determines a vehicle's energy consumption rate (gallons/mile, or kWh/mile or Btu/mile for electric and alternative-fueled vehicles). This value is multiplied by national average emission factors for the various pollutants to give emission rates in grams per mile. The greenhouse gas emissions portion of these results is shown in our master tables as the GHG number, given in tons per year.

Vehicle weight is used as the basis for estimating manufacturing impacts. Standardized, model-specific data on the environmental damage of vehicle manufacturing are not available. Therefore we use average manufacturing-sector emission factors and average breakdowns of vehicle materials by weight. These statistics determine the average emissions of each pollutant per unit of vehicle weight, which are multiplied by vehicle weight and divided by average vehicle lifetime mileage to obtain values of grams per mile of emissions related to manufacturing. We did not have sufficient data to estimate vehicle disposal and scrappage impacts, but these impacts are much smaller than manufacturing and in-use impacts, and, in general, would also be proportional to vehicle weight.

Having determined the average emission rates for each major stage of the vehicle's lifecycle (including those associated with the fuel consumed), the next step is to determine the relative environmental damage done by each pollutant. An economics-based approach for assessing environmental harm involves estimates of *damage costs* associated with a given pollutant. Specified, for example, in cents per gram (¢/g) of pollutant, these estimates reflect the costs to society of illnesses and premature deaths associated with pollution. Damage cost estimation involves uncertainties, of course, but it may also fail to reflect the full value we place on our health, environmental quality, and the protection of ecosystems. In spite of these limitations, damage costs provide a rational and consistent way to account for the different effects of various pollutants, and so we apply them to the emissions rates calculated from the vehicle data.

It is very difficult to estimate a damage cost for CO_2 and other greenhouse gases. The damage due to global warming is just beginning to occur and the worst risks are largely in the future. Therefore, we cannot look back at the harm that has already occurred—as has been done for conventionally regulated pollutants such as NO_x and PM—in order to estimate damage costs. However, because of the grave risks and growing concerns about greenhouse gas emissions, we give global warming concerns equal weight to other forms of air pollution in determining our green vehicle ratings. Therefore, we assigned CO_2 emissions a cost value such that, for the average 1999 light duty vehicle, half of the overall environmental harm is associated with global warming risks and the other half is associated with the health effects of conventional air pollutants.

Multiplying the gram-per-mile pollutant rates by their appropriate cents-per-gram damage costs (which vary by pollutant and location of emissions) yields environmental impact estimates in cents per mile (¢/mi). For conventionally regulated pollutants, adding these estimates up for a typical year of driving results in the "Health Cost" number shown in the main tables.

Adding up the ¢/mi estimates for all pollutants, including greenhouse gases, gives a total impact estimate for a given vehicle, which we term its *environmental damage index* (EDX). The EDX is the main result of our analysis for each vehicle and it provides the common metric with which we compare different makes and models. The EDX represents environmental harm; thus, the lower the EDX, the greener the vehicle.

For a green scoring system, greener vehicles should get higher scores. Therefore, we converted the EDX to the Green Score on a scale of 0–100 by grading along a curve, using a mathematical function specified so that an EDX of zero corresponds to a Green Score of 100. Finally, to determine the class ranking symbols, we examined the range of EDX values within each vehicle class. Cutpoints were determined on the basis of the distribution unique to each class. In addition, for a model to earn a checkmark (✔), it has to have an EDX that is lower than the overall average as well as being among the lowest in its class.

Further information on the analysis behind the ratings may be obtained by ordering our technical report, *Rating the Environmental Impacts of Motor Vehicles: The Green Guide to Cars and Trucks Metholology*, available from ACEEE publications.

List of Acronyms

Throughout the Green Guide to Cars and Trucks we've introduced and used a variety of acronyms and abbreviations. They are listed here along with their definitions to make it easier to follow the text.

ACEEE	American Council for an Energy-Efficient Economy
Btu	British Thermal Unit: a measure of energy
CARB	California Air Resources Board
CO	Carbon monoxide: a hazardous tailpipe pollutant
CO_2	Carbon dioxide: a product of fuel combustion and the principal greenhouse gas that causes global warming
DOE	U.S. Department of Energy
EDX	Environmental damage index
EPA	U.S. Environmental Protection Agency
FFV	Flexible Fuel Vehicle
GHG	Greenhouse gas: causes global warming
HC	Hydrocarbon: toxic and ozone-forming pollutant
kWh	kilowatt hour: a measure of electrical energy
LEV	Low Emission Vehicle
MPG	Miles Per Gallon: a measure of fuel economy
NHTSA	National Highway Traffic Safety Administration
NO_x	Nitrogen oxides: pollutant of multiple hazards
SO_2	Sulfur dioxide: pollutant of multiple hazards
SULEV	Super Ultra Low Emission Vehicle
Tier 1	Principal Federal emission standard since 1994
TLEV	Transitional Low Emission Vehicle
ULEV	Ultra Low Emission Vehicle
ZEV	Zero Emission Vehicle

Bibliography

American Automobile Association. 1997. Gas Saving Tips. Internet document: http://www.theautochannel.com/content/news/press/date/19971106/press007836.html

Bradsher, K. 1997. A Deadly Highway Mismatch Ignored. *The New York Times*, p. A1, September 24.

Calvert, J.G., J.B. Heywood, R.F. Sawyer, and J.H. Seinfeld. 1993. Achieving Acceptable Air Quality: Some Reflections on Controlling Vehicle Emissions. *Science*, 261, pp. 37–45.

Consumer Reports. 1996. Turning Up the Heat, Special Report on Global Warming and Environment. *Consumer Reports*, pp. 38–44, September.

———. 1997. Clearing the Air: Is Our Air Clean Enough? Special Report on Air Quality. *Consumer Reports*, pp. 36–38, August.

———. 1997. How Safe Are Sport Utility Vehicles? *Consumer Reports*, p. 54, June.

———. 1998. *New Car Yearbook 1999*. Yonkers, N.Y.: Consumers Union of U.S., Inc.

DeCicco, J., and M. Ross. 1994. Improving Automotive Efficiency. *Scientific American*, 271 (6), pp. 52–5.

DeCicco, J., and M. Thomas. 1999. *Rating the Environmental Impacts of Motor Vehicles: The Green Guide to Cars and Trucks Methodology*. Technical Report. Washington, D.C.: American Council for an Energy-Efficient Economy.

Delucchi, M.A. 1991. *Emissions of Greenhouse Gases from the Use of Transportation Fuels and Electricity*. Report ANL/ESD/TM-22. Argonne, Ill.: Argonne National Laboratory, Center for Transportation Research.

———. 1996. *The Annualized Social Costs of Motor Vehicle Use in the U.S., 1990–1991: Summary of Theory, Data, Methods, and Results*. Report 1. Davis, Calif.: Institute of Transportation Studies, University of California.

Dyson, C., B. Magavern, J. Shah, and T. Stephens. 1995. *The Green Buyer's Car Book*. Washington, D.C.: Public Citizen.

Energy Innovations. 1997. *Energy Innovations: A Prosperous Path to a Clean Environment*. Washington, D.C.: Alliance to Save Energy,

American Council for an Energy-Efficient Economy, Natural Resources Defense Council, Tellus Institute, and Union of Concerned Scientists.

Gillis, J. 1998. *The Car Book*. New York: Harper Collins.

———. 1998. *The Truck, Van, and 4x4 Book*. New York: Harper Collins.

Greene, D.L. 1997. Economic Scarcity: Forget Geology, Beware Monopoly. In Special Issue, "Running on Empty: Emerging Challenges in Global Energy Security." *Harvard International Review*, XIX (3), pp. 16–19; 65–67, Summer.

Hwang, R. 1997. *Are Cars Still a Problem? Real-World Emission Reductions from Passenger Vehicles over the Past 30 Years*. Berkeley, Calif.: Union of Concerned Scientists.

Keoleian, G., K. Kar, M.M. Manion, and J.W. Buckley. 1997. *Industrial Ecology of the Automobile: A Lifecycle Perspective*. Warrendale, Penn.: Society of Automotive Engineers.

Makower, J. 1992. *The Green Commuter*. Washington, D.C.: National Press Books.

Ross, M., R. Goodwin, R. Watkins, M. Wang, and T. Wenzel. 1995. *Real-World Emissions from Model Year 1993, 2000 and 2010 Passenger Cars*. Washington, D.C.: American Council for an Energy-Efficient Economy.

Schipper, L., S. Meyers, R.B. Howarth, and R. Steiner. 1993. *Energy Efficiency and Human Activity: Past Trends, Future Prospects*. Cambridge, U.K.: Cambridge University Press.

Seidel, S.R. 1996. Keeping Cars Cool. Chapter 11 in *Ozone Protection in the United States: Elements of Success*. Ed. E. Cook. Washington, D.C.: World Resources Institute.

Shprentz, D.S. 1996. *Breath-Taking: Premature Mortality Due to Particulate Air Pollution in 239 American Cities*. Washington D.C.: Natural Resources Defense Council.

U.S. Environmental Protection Agency. 1992. *What You Can Do to Reduce Air Pollution*. EPA 450-K-002. Washington, D.C.: EPA.

———. 1994. *Tips to Save Gas and Improve Mileage*. EPA Fact Sheet OMS-17 (EPA 400-F-92-004). Washington, D.C.: EPA.

———. 1994. *Your Car and Clean Air: What You Can Do to Reduce Pollution*. EPA Fact Sheet OMS-18 (EPA 400-F-93-002). Washington, D.C.: EPA.

———. 1995. *National Air Quality and Emissions Trends Report, 1994*. EPA 454-R-95-014. Research Triangle Park, N.C.: EPA.

Yergin, D. 1991. *The Prize: The Epic Quest for Oil, Money, and Power*. New York: Simon and Schuster.

About the Authors

JOHN DECICCO is a Senior Associate and the Transportation Program Director for ACEEE. A leading expert on transportation energy use and emissions, he conducts research on automotive technologies and has published extensively, writing for both popular and technical audiences. His expertise includes motor vehicle efficiency, energy use and greenhouse gas emissions of transportation systems, as well their other environmental and economic impacts. He tracks the status of new and emerging technologies, and analyzes programs and policies for stimulating the adoption of technologies that offer environmental and economic benefits. He has a long-standing interest in developing marketing-oriented approaches and better consumer information on environmental issues, and pioneered the *Green Guide to Cars and Trucks* in order to provide such information for the automotive marketplace. He is also coeditor of the book *Transportation, Energy, and Environment: How Far Can Technology Take Us?* and the study *Energy Innovations: A Prosperous Path to a Clean Environment*. He received his Ph.D. in Mechanical Engineering from Princeton University in 1988, where he was affiliated with the University's Center for Energy and Environmental Studies.

MARTIN THOMAS is a Research Assistant at ACEEE, where he works on transportation energy issues. His work includes modeling of vehicle fuel economy and emissions and life cycle analysis of the environmental impacts of automobiles. Before joining ACEEE, he worked as a research assistant in the analysis of vehicle emissions at Los Alamos National Laboratory and at the University of Michigan. He holds an M.S. in physics from the University of Michigan. He drives a zero-emission 12-speed bicycle.

ALSO AVAILABLE FROM ACEEE

Books on Transportation and Automotive Issues

Transportation, Energy, and Environment: How Far Can Technology Take Us? examines the potential for technological improvements in highway vehicles, railroads, and aircraft to move the U.S. transportation system toward greater sustainability.

Edited by John DeCicco, ACEEE, and Mark Delucchi, Institute of Transportation Studies.

ISBN 0-918249-28-7 • Soft cover, 6"x 9", 278 pp., index, 1997, $33

Transportation and Energy: Strategies for a Sustainable Transportation System analyzes how transportation energy choices made by citizens, policymakers, and planners will affect national goals of mobility, accessibility, environmental quality, economic growth, and energy security.

Edited by Daniel Sperling and Susan A. Shaheen, Institute of Transportation Studies.

ISBN 0-918249-20-1 • Soft cover, 6"x 9", 320 pp., index, 1995, $31

Transportation and Global Climate Change focuses on the transportation sector's role in global warming and what can be done about it.

Edited by Danilo J. Santini, Argonne National Laboratory, and David L. Greene, Oak Ridge National Laboratory

ISBN 0-918249-17-1 • Soft cover, 6"x 9", 357 pp., index, 1993, $31

Consumer Guides

Consumer Guide to Home Energy Savings

This illustrated, 267-page book contains expanded listings of energy-efficient appliances and practical suggestions on how to reduce energy use and help protect the environment through home improvements. This guide is available in bookstores ($8.95) or directly from ACEEE ($13.95 postpaid).

Guide to Energy-Efficient Office Equipment

This second edition can help office equipment buyers select and use efficient equipment. It includes information on the energy use characteristics of personal computers, monitors, printers, copiers, facsimile machines, combination fax/printer/scanners, and retrofitting equipment with power-management devices. The relationship between energy consumption and paper use in the workplace is examined and guidance on public-access databases is provided to obtain product-specific information ($17 postpaid).

Guide to Energy-Efficient Commercial Equipment

Using this guide, buyers can specify and select the most energy-efficient commercial equipment—lighting options, space conditioning equipment, and motors for commercial applications. It presents criteria for selecting efficient equipment and lists the most efficient commercial equipment on the market, as well as provides information about how to operate and maintain the equipment ($30.00 postpaid).

For More Information or To Order, contact ACEEE Publications Office

Phone: 202-429-0063 / Fax: 202-429-0193
ace3pubs@ix.netcom.com

Visit ACEEE's World Wide Web home page at http://aceee.org